SpringerBriefs in Regional Science

SpringerBriefs present concise summaries of cutting-edge research and practical applications across a wide spectrum of fields. Featuring compact, authored volumes of 50 to 125 pages, the series covers a range of content from professional to academic. SpringerBriefs in Regional Science showcase emerging theory, empirical research and practical application, lecture notes and reviews in spatial and regional science from a global author community.

More information about this series at http://www.springer.com/series/10096

Rita De Siano · Valerio Leone Sciabolazza ·
Alessandro Sapio

Regional Resilience to Climate and Environmental Shocks

A Spatial Econometric Perspective

Springer

Rita De Siano
Parthenope University of Naples
Naples, Italy

Valerio Leone Sciabolazza
Parthenope University of Naples
Naples, Italy

Alessandro Sapio
Parthenope University of Naples
Naples, Italy

ISSN 2192-0427 ISSN 2192-0435 (electronic)
SpringerBriefs in Regional Science
ISBN 978-3-030-54587-1 ISBN 978-3-030-54588-8 (eBook)
https://doi.org/10.1007/978-3-030-54588-8

This Springer imprint is published by the registered company Springer Nature Switzerland AG
The registered company address is: Gewerbestrasse 11, 6330 Cham, Switzerland

Acknowledgements

We are grateful to Johannes Glaeser, Editor of Economics & Political Science for Springer Heidelberg, and to the Editors of the series "SpringerBriefs in Regional Science" for carefully reading our manuscript and providing constructive and insightful comments. We thank Prof. Roberto Basile (University of L'Aquila) and Simon Hailstone (England NHS) whose lectures inspired part of the R scripts presented in this book. Financial support from Parthenope University of Naples, *Bando di Ateneo per il sostegno alla partecipazione ai bandi di ricerca competitiva per il triennio 2016–2018*, is gratefully acknowledged.

Contents

1 Introduction 1
References 5

2 Economic Resilience and Regional Disparities: The Value Added of Spatial Analysis 7
2.1 Origin and Evolution of the Concept of Economic Resilience 7
2.2 Regional Economic Resilience 9
2.2.1 Engineering, Ecological and Evolutionary Resilience 9
2.3 Regional Economic Resilience to the Great Recession 13
2.4 Regional Resilience to Climate-Related Emergencies and Disasters 15
2.5 Resilience Measurement and Empirical Strategies 19
2.6 A Prime Overview of Spatial Dependence in Regional Economic Resilience: The Added Value of Spatial Analysis 22
References 24

3 Spatial Econometric Models: Theory 31
3.1 Introduction to Spatial Analysis 31
3.2 Modelling Space 33
3.3 Exploratory Spatial Data Analysis 34
3.4 Spatial Econometrics Models 36
3.4.1 Cross-Section Models 37
3.4.2 Panel Data Models 39
3.4.3 Specification Tests 40
3.4.4 Estimation Approaches 41
References 41

4 A Tutorial on Modelling Geographic, Economic and Social Interactions Using GIS Methods with R 45
4.1 Introduction 45
4.2 Chapter Overview 47

4.3 Load Data and Software Packages 47
4.4 Working with Coordinates 49
4.5 Modelling Spatial Interactions 50
4.5.1 Contiguity-Based Adjacency Matrices 50
4.5.2 Distance-Based Adjacency Matrices 53
4.5.3 K-Nearest-Neighbour Weights Matrix 54
4.6 Modelling Social Interactions 57
4.6.1 Creating and Plotting an Edgelist 57
4.6.2 Creating and Plotting Network Data 60
4.7 Spatial Tests for Autocorrelation 63
4.7.1 The Join Count Test 65
4.7.2 The Global Moran's *I* Test 65
4.7.3 The Global Geary's *c* Test 66
4.7.4 Local Moran's *I* Test 67
4.8 Conclusions 69
References 70

5 Resilience to Climate Change: Spatial Ricardian Analysis 73
5.1 Introduction 73
5.2 Measurement Issues 76
5.3 Spatial Ricardian Models 78
5.4 Estimates of Spatial Climate Spillovers on Land Value 80
5.5 Outlook and Conclusion 82
5.5.1 Resilience, Adaptation and Mitigation 82
5.5.2 Econometric and Simulative Impact Assessment: From Competition to Cooperation 83
References 86

6 Resilience to Climate Impacts and Spatial Propagation in the Power Industry 89
6.1 Introduction 89
6.2 Econometric Specification Issues 92
6.3 Power Demand and the Resilience of Power Systems 94
6.3.1 Resilience of a Power Market to Structural Change in Power Demand 95
6.3.2 Literature Review 99
6.4 The Adoption of Renewable Energy Technologies 102
6.4.1 Theoretical Insights 102
6.4.2 Literature Review 104
6.5 Conclusion 106
References 107

7 Conclusion and Open Issues 111
References 114

Chapter 1
Introduction

Abstract This introductory chapter provides a chapter-wise description of the book. It illustrates the value added of critically reviewing the state of the art on the economic resilience of regions with respect to shocks that are related to climate change and environmental issues. Insights are also offered on the key role of spatial econometrics.

Keywords Resilience · Regions · Climate shocks · Spatial econometrics

In the face of disruptive climate change, knowledge stemming from the economic and econometric analysis of spatial propagation processes can provide fundamental insights on the resilience of economic systems at various aggregation scales (countries, regions) and on the design of climate policies, with a view to stronger adaptation and mitigation capabilities of our economic systems. Climate change is a global externality, yet it hits economic systems by triggering local shocks that affect capital stocks, physical infrastructures and labour productivity, thereby hindering the ability of regions and countries to produce, to trade and to reallocate unemployed workers. There is a blossoming literature on spatial econometric assessments of economic resilience, with respect to various types of shocks (recessions, policy decisions, climate change, natural disasters).

Several considerations motivate this book within the broad context of the current policy framework on environmental sustainability. The vulnerability of countries/regions or societies to the effects of external shocks, such as environmental changes, depends not only on the magnitude of the shocks they experience but also on their capacity to adapt to or face such stresses. This ability is identified in the literature with the term of resilience. Resilience has been measured with reference to several macroeconomic variables, at the national and regional resolutions, such as employment, wages and international trade links. Highly relevant applications of spatial econometrics have been recently published in order to assess the effects of recessionary shocks (Fingleton et al. 2012) or of Brexit (Fingleton 2019), shedding

R. De Siano et al., *Regional Resilience to Climate and Environmental Shocks*, SpringerBriefs in Regional Science, https://doi.org/10.1007/978-3-030-54588-8_1

light on the advantages of such models with respect to non-spatial alternatives (Martin and Gardiner 2019). Closer to our interests are the assessments of resilience to natural disasters (Lima and Barbosa 2019; Muñoz and Tate 2016), whose frequency correlates with climate change. The use of spatial analytical tools, such as those provided by spatial econometric methodologies (Anselin 1988; LeSage and Pace 2009), enables to account for the presence of spatial effects that otherwise would lead to an incorrect representation and understanding of the true causal processes at work. Nevertheless, even when shocks produce global externalities as in the case of climate change, it becomes crucial to account for the presence of spatial interdependences that relate spatial units directly, as with the effects of climate-related natural disasters and infrastructure disruptions. Not only the shocks themselves but also environmental policy strategies may produce effects that spread beyond the geographical boundaries generating beneficial or harmful externalities on the neighbouring regions.

To our knowledge, although the literatures on economic resilience, climate shocks and spatial econometrics are increasingly overlapping and inter-breeding, there is no systematic treatment of their relationships. Books on the theory of spatial econometrics illustrate novel solutions to estimation and testing issues, but most of those methods have not yet been applied to resilience issues. Other spatial econometric books explore traditional economic problems in order to highlight the advantages of spatial models over non-spatial alternatives. Contributions closer to our proposal are included in Duranton et al. (2015), where some chapters deal with the relationship between cities and the environment. The book by Pinto et al. (2018) tackles the economic resilience of regions through various methodological angles, including a chapter by Bruckmeier and Pires on a socio-ecological perspective on resilience and sustainability (Bruckmeier and Pires 2018). Hallegatte's (2016) book on natural disasters and climate change does not take an explicitly spatial perspective, but its theoretical and empirical discussions indicate space as an important mean for shock propagation, stressing the need to estimate indirect economic consequences and regional redistribution effects.

Our ambition is to collect and critically review the existing results on the economic resilience of regions with respect to shocks that are related to climate change and environmental problems, thereby offering a comprehensive state of the art on the subject. Post-graduate students and researchers may benefit from this endeavour also because the book will illustrate the theory of spatial econometrics and provide practical guidance for potentially new applications.

In what follows, we provide a chapter-wise description of the book, through summaries of their contents.

Chapter 2 offers a theoretical detour on the concepts of economic resilience in general. The definitions of resilience and the ways of measuring it are discussed in depth (see also the reviews in Toth 2015; Turner 2010). Based on the existing research approaches, the chapter argues that economic, political and environmental shocks may strongly influence the evolution of a geographical area. A resilient region is able to respond positively to changes and is able to maintain its core functions despite those stresses. An important issue is also understanding what enables a geographical area or a society to bounce back. The capacity to react can be influenced by various factors such as economic structure, dynamism, market and political conditions, resources endowments and the propensity to innovate. The economic debate, indeed, has focused on what usually makes a region most resilient in order to drive policy-makers towards decisions that reduce the vulnerability of regions/societies to shocks and enhance their ability to better respond to and recover from a shock.

Along with being exposed to the theory of economic resilience, readers interested in how different regions react to external shocks would highly benefit from learning the basics of spatial econometrics, for the reasons outlined in this introductory chapter. This is the subject of Chap. 3. The chapter starts from the basic description of the Exploratory Spatial Data Analysis (ESDA), providing global and local tests for the presence of spatial patterns highlighting a clustering in the space of high or low values for regional indicators of resilience. The chapter then illustrates the main models of spatial interaction (Anselin 1988): depending on the type of interaction between observations of neighbouring units, different specifications may be followed (SAR, SEM, SDM, SARAR and so forth) according to the results of model selection tests (usually likelihood-based tests).

Chapter 4 is dedicated to the practical application of spatial econometrics. The aim of this chapter is to acquaint the reader in a self-contained manner with a number of analytical tools to examine the structure of social and spatial relations and to investigate the impact of these relations on economic outcomes. The complex web of relations existing among economic agents is formalized through the use of graph theory and the spectral analysis of matrices. This approach conveys relevant information on the position of the agent in the social and geographical space and provides the means for a theoretically founded guide to analyse individual economic behaviours inside the structure of agents' interactions. Specifically, the focus is put on the identification of potential sources of externalities shaping individual actions, and on the choice of the appropriate statistical modelling for the empirical analysis. The methodologies discussed are taken from both spatial statistics and network analysis. These are used to define and map the distance between agents in different ways, e.g. physical distance, which is represented by a geometric system that has Cartesian coordinates and social distance, which is used to operationalize how social and economic ties are formed. The formalization of distance is used to identify agents' neighbourhood, that is, the potential source of spillover effects. In particular, this chapter introduces a number of descriptive statistics to test for the presence of autocorrelation in the composition of the agent's neighbourhood and assesses the level of interaction between the social and the geographic space in which agents are embedded. The result of this analysis is meant to provide significant information on the potential presence of factors that are crucial to the understanding of the causal relation between individual behaviour and the resulting economic outcome. Because of this, it may suggest the choice of a specific empirical strategy to investigate the behaviour of economic agents in several fields. This chapter adopts a hands-on approach. For this reason, the reader is walked through the discussion of the various methodologies using an application to trade flows in Africa together with replicable examples in R.

International trade represents a major channel of externalities among many sectors of the national economies, and it contributes to determine the level of vulnerability of a country or region. As such, it motivates its analysis in the context of the study of economic resilience.

Finally, Chaps. 5 and 6 review the existing econometric results on some aspects of economic resilience to climate and environmental shocks. Chapter 5 begins with a characterization of resilience to shocks due to climate change, following Bahadur et al. (2013) as well as Cannon and Müller-Mahn (2010), and relates it to the climate econometrics literature, reviewed in Hsiang (2016) and Dell et al. (2014). Focusing on the case of agriculture, the chapter, in particular, illustrates the Ricardian approach and how it has been extended to account for spatial autocorrelation in economic outcomes (i.e. farm or land values) or in weather variables. The chapter then proceeds to illustrate the existing applications of spatial econometrics in the analysis of spatial spillovers generated by climate shocks. In particular, considerable interest in the resilience of rural areas and productions to weather extremes is testified by a number of published articles, some of which we review in more detail. Advantages and limitations of spatial econometric assessments are discussed in light of the types of adaptation strategies that are central in climate policy and in view of possible integration between climate econometrics and simulation models.

In Chap. 6, the focus is on the electric power sector and on the adoption of renewable energy technologies. This is a highly relevant issue in performing a comprehensive assessment of a country's resilience to climate shocks. Indeed, an increasing penetration of locally available renewable energy sources allows a country to reduce its dependence on energy imports. To the extent that shocks are transmitted through international trade links, a lower dependence on energy imports may isolate a country/region from the shocks and may make economic resilience easier to achieve. In policy-making terms, spatial dependencies among investments in new renewable energy plants are worth taking into consideration in the design of support policies. The chapter recollects the role of space in the structure and dynamics of the energy industry under different technological paradigms, following the seminal work by Nijkamp (1983), the research on nodal pricing by Bohn et al. (1984), and the comparison between distributed and centralized generation paradigms performed by Künneke (2008). Next, the chapter reviews the applications of spatial econometrics to the diffusion of photovoltaic panels and to the resilience of electricity demand to shocks that may affect distribution and transmission networks (e.g. Maliszewski and Perrings 2012). With regard to both case studies, literature reviews are preceded by theoretical outlines that help elucidate the interpretation of the spatial econometric estimates as well as their potential for improvements. Implications for the robustness of the industrial structure to shocks are drawn.

Concluding remarks and a discussion of open research issues are included in Chap. 7.

References

Anselin, L. (1988). *Spatial Econometrics: Methods and Models*. Dordrecht, The Netherlands: Kluwer Academic Publishers.

Bahadur, A. V., Ibrahim, M., & Tanner, T. (2013). Characterising resilience: Unpacking the concept for tackling climate change and development. *Climate and Development, 5*(1), 55–65.

Bohn, R. E., Caramanis, M. C., & Schweppe, F. C. (1984). Optimal pricing in electrical networks over space and time. *Rand Journal of Economics., 15*(3), 360–376.

Bruckmeier, K., & Pires, I. (2018). Innovation as transformation: Integrating the socio-ecological perspectives of resilience and sustainability. *Resilience and regional dynamics* (pp. 209–231). Cham: Springer.

Cannon, T., & Müller-Mahn, D. (2010). Vulnerability, resilience and development discourses in context of climate change. *Natural Hazards, 55*(3), 621–635.

Dell, Melissa, Jones, Benjamin F., & Olken, Benjamin A. (2014). What do we learn from the weather? The new climate-economy literature. *Journal of Economic Literature, 52*(3), 740–798.

Duranton, G., Henderson, V., & Strange, W. (Eds.). (2015). Handbook of regional and urban economics. Elsevier.

Fingleton, B., Garretsen, H., & Martin, R. L. (2012). Recessionary shocks and regional employment. *Journal of Regional Science, 52*(1), 109–133.

Fingleton, B. (2019). Exploring Brexit with dynamic spatial panel models: Some possible outcomes for employment across the EU regions. *The Annals of Regional Science*, 1–37.

Hallegatte, S. (2016). Natural disasters and climate change. *Springer*,. International.

Hsiang, S. (2016). Climate econometrics. *Annual Review of Resource Economics, 8*, 43–75.

Künneke, R. W. (2008). Institutional reform and technological practice: The case of electricity. *Industrial and Corporate Change, 17*(2), 233–265.

LeSage, J. P., & Pace, R. K. (2009). *Introduction to spatial econometrics*. New York: CRC Press.

Lima, R. C. D. A., & Barbosa, A. V. B. (2019). Natural disasters, economic growth and spatial spillovers: Evidence from a flash flood in Brazil. *Papers in Regional Science, 98*(2), 905–924.

Maliszewski, P. J., & Perrings, C. (2012). Factors in the resilience of electrical power distribution infrastructures. *Applied Geography, 32*(2), 668–679.

Martin, R., & Gardiner, B. (2019). The resilience of cities to economic shocks: A tale of four recessions (and the challenge of Brexit). *Papers in Regional Science, 98*, 1801–1832.

Muñoz, C., & Tate, E. (2016). Unequal recovery? Federal resource distribution after a Midwest flood disaster. *International Journal of Environmental Research and Public Health, 13*(5), 507.

Nijkamp, P. (1983). Regional dimensions of energy scarcity. *Environment and Planning C: Government & Policy, 1*(2), 179–192.

Pinto, H, Noronha, T, Vaz, E. (Eds.) (2018) Resilience and regional dynamics: An international approach to a new research agenda, Springer.

Toth, B. I. (2015). Regional economic resilience: concepts, empirics and a critical review. *Miscellanea Geographica, 19*(3), 70–75.

Turner Ii, B. L. (2010). Vulnerability and resilience: Coalescing or paralleling approaches for sustainability science? *Global Environmental Change, 20*(4), 570–576.

Chapter 2
Economic Resilience and Regional Disparities: The Value Added of Spatial Analysis

Abstract The idea of regional economic resilience refers to the ability of regions to resist and recover from a given shock. The chapter offers an overview of the definitions of resilience usually found in the literature and the ways of measuring when dealing with regional economics. Actually, the concept of resilience may refer to various dimensions of regional economic performances, such as the vulnerability or the sensitivity to different types of shocks, the resistance to the impact of economic shock impacts, the way firms, workers and institutions respond or adapt to shocks and, finally, the nature of recovery. The economic debated focused mainly on the research of the characteristics that would make each region more resilient, in order to drive policy-makers in building appropriate measures and strategies aimed at reducing the vulnerability of spatial systems to shocks and enhance their ability to better respond to and recover from the crises. However, various theoretical analyses evidence also the presence of beneficial or harmful externalities driven by policies in a given region on its neighbouring regions. To account for the presence of externalities and spillover mechanisms, empirical analyses may relate to spatial econometrics approaches that enable to consider both geographical and socio-economic proximities.

Keywords Regional economics · Economic resilience · Engineering resilience · Ecological resilience · Adaptive resilience · Resistance · Recoverability · Measures of resilience · Externalities · Spatial analysis

2.1 Origin and Evolution of the Concept of Economic Resilience

The Latin root of the term resilience, "resiliere", means "to leap back, to rebound or to shrink back again". As a consequence, the meanings most frequently found refer to two different features of systems present in nature that recall exactly the Latin origin of the word. The first is "toughness", that is, "the capacity to recover quickly from difficulties", applying to general responses of a system to different kinds of disturbances. The second is "elasticity", which has a stricter application to scientific

R. De Siano et al., *Regional Resilience to Climate and Environmental Shocks*, SpringerBriefs in Regional Science, https://doi.org/10.1007/978-3-030-54588-8_2

contexts and refers to "the ability of a substance or an object to spring back into shape".

Abstracting from the notion of resilience used in hard sciences, the term has had since the beginning a large application in the psychological (Werner 1977; Werner and Smith 1982, 2001; Gabriel 2005; Mergenthaler 2012) and ecological literatures (Holling 1973; Brand et al. 2011; Günther 2009; Gunderson et al. 2010; Gunderson and Holling 2002). More recently this concept has spread to the socio-economic disciplines. The main reason of such a diffusion is that "resilience thinking" offers a new perspective in addressing increasingly crucial issues for communities, institutions and economies, such as understanding the growing vulnerability of economic systems to sudden and unpredictably negative external shocks, and identifying the factors that enhance the capacity of economic actors and policy-makers to build successful responses.

An increasing interest in resilience has found currency also in the political debate (Rodin 2015; Zolli and Healy 2012). Most developed countries, indeed, have been characterized by epochal changes, in turn, caused by internal and external factors. The list of these causes can be vast, but climate change, globalization and automation rank among the most important game changers, by transforming production systems and business models. In particular, climate change is causing an intensification of natural disasters and the introduction of decarbonization processes in the energy sector. Moreover, climate change triggers migration flows that raise integration problems in the host countries, affected by the demographic transition. Finally, deep financial and economic crises have motivated the implementation of austerity and unorthodox monetary policies. In this changing scenario, practitioners in policy planning acknowledge that anticipating a potential adverse event can be crucial for a quick perception of the risks, in order to develop appropriate responses that allow to absorb or, even better, to prevent changes in the social and economic contexts (Giacometti and Teräs 2019; Bonß 2016). Eraydin (2012) defines resilience as the basis for a new planning paradigm and Walker et al. (2004) identify four "crucial aspects of resilience" growing in importance in the policy-making process. First, the capacity of a system to work under pressure, facing new external changes without losing the ability to recover; second, resistance, that is the ability to maintain core features despite the necessity of changes to recreate the previous state of normality; precariousness, which refers to the current trajectory of the system and the distance from a "threshold" which, if infringed, makes recovery difficult or impossible; finally, panarchy which refers to the influences from states and dynamics at scales above and below a system may undergo (Walker et al. 2004). Accordingly, resilience should be regarded as the potential of a system to reorganize while undergoing changes with the aim of keeping its configuration and functions. The lack of this potential determines the level of vulnerability of a system since, in addition to the magnitude of a shock, it depends on the capacity to adapt itself to or face a sudden change. An even taller challenge is raised by climate change, as climate-related shocks are ultimately anthropogenic in nature, thereby posing the additional question of identifying effective mitigation policies. Yet the climate-economy and the economy-climate chains of causation manifest themselves over different time scales.

2.2 Regional Economic Resilience

In recent times, the concept of resilience has started to be used in regional science and spatial planning, while acquiring new meanings and relevance. The evolutionary perspective in economic geography and the acknowledgement that shocks may have a transitory as well as a persistent effect on growth paths (Blanchard and Katz 1992; Cerra and Saxena 2008; Martin 2012; Simmie and Martin 2010) contributed to apply the resilience thinking approach also to regional economic issues. This interest, both in urban and regional analyses, was triggered by the outbreak of the global economic and financial crisis that started in the United States in 2008. The increasing perception of uncertainty highlighted the need to understand why some countries, regions or cities respond differently to economic downturns. The search for the conditions and features that favours regional economic resilience the most became crucial (Angulo et al. 2018; Boschma 2015; Bristow 2010; Bristow and Healy 2018, Cambridge Journal of Regions, Economy and Society 2010; Di Caro and Fratesi 2018; Fingleton et al. 2012, 2015; Hassink 2010; Hudson 2010; Pike et al. 2010; Martin 2012, 2018; Martin and Gardiner 2019; Martin and Sunley 2015; Sensier et al. 2016; Simmie and Martin 2010; Webber et al. 2018). Besides, the globalization process, by contributing to make places and regions more permeable and more vulnerable to external shocks, gave inspiration to the earliest empirical studies on new paths of resilience allowing for the presence of spatial interdependencies (Fingleton and Palombi 2013; Doran and Fingleton 2016; Modica and Reggiani 2015).

2.2.1 *Engineering, Ecological and Evolutionary Resilience*

In his seminal work "Regional Economic Resilience, Hysteresis and Recessionary Shocks" (Martin 2012), Martin distinguished three different interpretations of resilience which gave rise to as many theoretical streams, namely, engineering, ecological and evolutionary (Martin 2012; Martin and Sunley 2015).

Engineering resilience refers to the stability of a system near an equilibrium or a steady state. As the system is supposed to be in equilibrium before the shock, empirical analyses are supposed to focus on the resistance and the response, as well as at the speed of recovery in returning to the pre-shock state (Holling 1973). For this to happen, it is necessary that a self-correcting mechanism automatically activates, in order to absorb the shock without changing the core productive structure and the socio-economic configuration of the system (Fig. 2.1). By contrast, the presence of market frictions may prevent the system from reacting promptly, causing a condition of lack of resilience (Fig. 2.2).

However, this approach has been criticized for its simplistic assumption of a single equilibrium towards which the system should bounce back after the shock. The idea, indeed, hardly reconciles with the possibility that a region in trouble can evolve

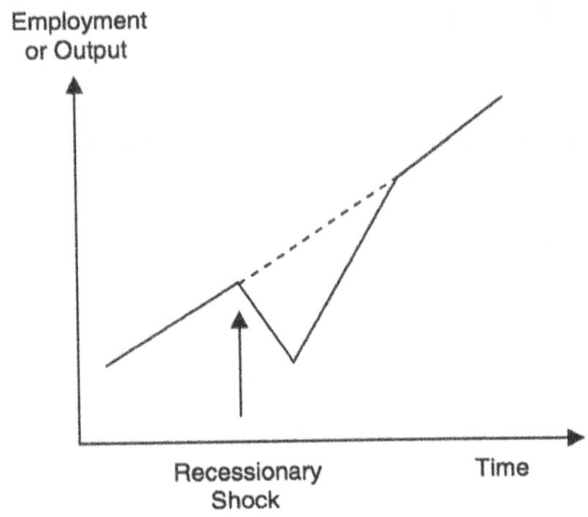

Fig. 2.1 Impact of a recessionary shock on a region's growth path: region returns to pre-shock growth trend

to a different growth path. The regional production system may in fact respond by removing unproductive activities and opening up to new profitable sectors leading the economy towards a different steady state with respect to the starting one (Bonß 2016; Muštra et al. 2016). The result may be the existence of multiple equilibria, with the economic system moving from one state to another. This phenomenon is explained by Martin (2012) using the concept of hysteresis, originally used in the field of natural sciences.

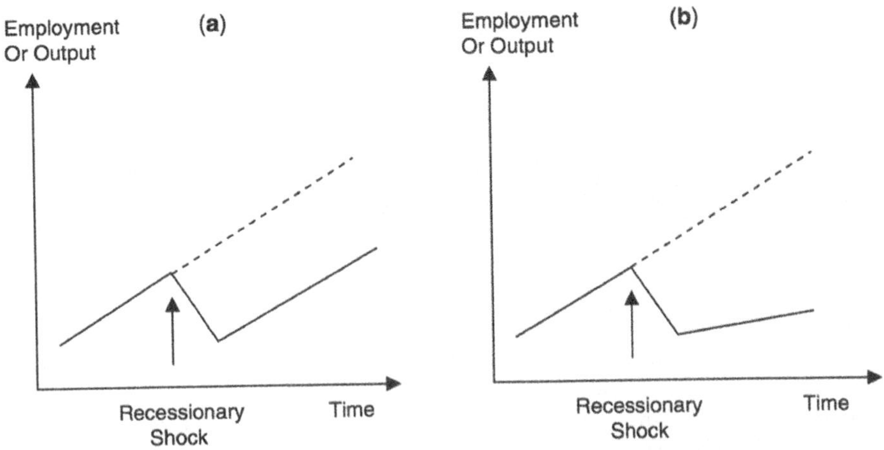

Fig. 2.2 Negative impacts of a recessionary shock on a region's growth path: (**a**) permanent decline in level, resumption of pre-recession growth rate, (**b**) permanent decline in level and lowered growth rate (*Source* Martin and Gardiner 2019)

Besides, some scholars (Martin 2010, 2012) consider that an economy should not be necessarily in equilibrium and, by contrast, it may lie on a relatively stable growth path. Under this circumstance, the self-correcting mechanism can be assimilated to the response of the economy to fluctuations as it is described in the "plucking model" (Friedman 1993). This model predicts that a disturbance, as in the case of a recession, exerts only a transitory effect on the economy which, soon or later, bounces back to its long-run growth path. Output and employment may, of course, fall after an economic contraction but their reaction will depend on the economy's own features in terms of resource endowments, potential for innovation, quality of institutions and policy priorities, local attractiveness and, as it has become increasingly clear in recent times, exposure to climate change impacts. This may result in a possible differentiation across the reactions of different economies.

Finally, it should be noted that a shock causing a structural change in the system may affect its own resilience potential to future shocks (Simmie and Martin 2010). If this is the case, then resilience may also evolve over time. Ecological resilience defines the scale of a disturbance a regional system is able to absorb before changing its structure and moving to a new equilibrium state. Therefore, in contrast to the engineering approach, the ecological concept of resilience assumes that a system may have multiple equilibria (Hassink 2010; Simmie and Martin 2010). Some scholars assert that economies never are in equilibrium, rather they are driven by knowledge which constantly changes (Ramlogan and Metcalfe 2006; Dosi and Nelson 1994). Systems are supposed to change constantly, but only when an unexpected event or a shock of a certain magnitude takes place does the adjustment become observable. The main aim of ecological resilience analysis is to define the magnitude of the shock that can be tolerated before the system changes. If the shock is too severe, its effects may be permanent: this is when hysteresis occurs (Setterfield 2010).

Systems happen to be more resilient the more they are flexible or able to move to new typologies of business by means of innovation and adaptation, as well as by mitigating the root causes of potentially adverse shocks. The state towards which a system moves may differ depending on initial conditions, on agents' capacity to predict the unexpected event and on the aftermaths of previous shocks. With respect to climate change, initial conditions seem to be all the more essential in determining resilience, as reactive measures adopted to cope with previous non-climatic shocks may not be informative under the future state of the climate, and most existing predictive techniques prove ineffective with respect to a highly non-linear and unprecedented dynamics that is deemed to unfold over a long time horizon. Individual agents, in particular, are even less equipped to predict future climate-related events.

After a shock, the system may move either towards a worse path, characterized by a permanent decline in the level of the main economic variables and/or a lower growth rate, or, on the opposite, it can revise productive capacity and experience a dramatic increase in unemployment. If this contraction is not accompanied by a selection of the best production activities and workers' capabilities, the economy can resume its previous growth rate but only on a lower trend path. In a more pathological case, the damage to the productive sector is so profound that the whole economy, not just least efficient firms and workers, may undergo a deep and permanent contraction preventing its recovery of the pre-shock growth rate. If this is the case, economies are regarded as low resilient. Such a danger is even more likely to materialize if one shifts the focus on a long-run, pervasive phenomenon such as climate change.

A deep recessionary shock, on the contrary, could exert also a positive effect on the economy. It may help to sweep unproductive and outdated activities and contribute to achieve a permanent increase in system performances by implementing favourable economic, political and institutional reforms (Martin and Sunley 2015; Caballero and Hammour 1994; Gali and Hammour 1993). The economy may reach a higher path characterized by better performances in terms of output levels or growth rates or both of them in comparison with the pre-shock situation. Economies exhibiting positive permanent effects of either type would be regarded as highly resilient (Fig. 2.3).

The last notion of resilience identified in the seminal work by Martin (2012) is adaptive resilience which is based on an evolutionary approach and denotes the ability of a region to reconfigure itself, that is, to adjust its structure (industries, firms, institutions, technologies) in response to shocks so as to keep itself on a sustainable growth path (Pendall et al. 2010; Pike et al. 2010). Resilience is a dynamic process of constant change emerging by decisions to leave a path in favour of a new one while keeping the capacity to allocate resources effectively. Martin et al. (2016) describe adaptive resilience as a multifaceted process comprising some key conditions: the exposure to risks by the main economic players (firms, industries, workers and institutions) and their resistance to the impact of economic shocks; the capacity to adjust in order to reorient their core activities and the degree of recoverability in terms of economic performances, including profitability, employment and investment. These conditions strongly depend on the magnitude, the nature and the duration of the shock on the one hand and, on the other, on local economic structures, resources

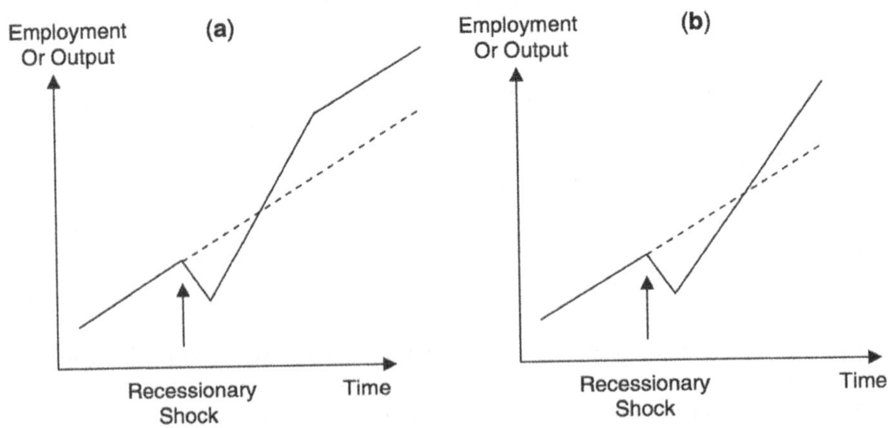

Fig. 2.3 Positive hysteretic impacts of a recessionary shock on a region's growth path: (**a**) recovery to higher level, resumption of pre-recession growth rate, (**b**) recovery to a sustained higher growth rate (*Source* Martin and Gardiner 2019)

and capabilities. Finally, the resistance to shocks may be strengthened also by the presence of an adequate framework of policies and programmes, at various scales, from supranational to national and local.

2.3 Regional Economic Resilience to the Great Recession

The Great Recession that begun in 2008 has provided the first major chance for the resilience approach to prove its usefulness in the field of regional economics. In fact, a host of empirical works on how different regions within a country react to economic shocks has been sparked by the Great Recession. After the global economic and financial crisis, some regions showed themselves more resilient as they resulted less affected by the severe crisis and/or they recovered more quickly than other regions. Local economies, in fact, are the result of a complex mixture of institutional, historical, economic and cultural aspects that result to be highly interdependent both over time and over space. The evolution of these aspects may change the regional ability to resist and to adapt to economic and environmental shocks, leading to a conceptualization of resilience that validates its evolutionary notion (Martin and Sunley 2015). A higher capacity of reaction may help a region to shift upwards its growth path and go beyond its pre-shock feasible "ceiling" in output or employment. This represents the situation in which a region exhibits a high degree of robustness and resilience.

From an empirical perspective, it is by now well established that regions differ in their resilience in terms of regional growth or employment. Fingleton et al. (2012, 2015), for example, found considerable differences in regional resilience as measured by employment patterns for the UK and EMU regions, respectively. Simmie and Martin (2010), by using narrower local units, find significant effects of economic resilience on regional disparities. The authors used a two cities case study, considering Cambridge and Swansea as units of analysis, to explore the usefulness of the adaptive cycle model in shaping regional economic resilience. These case studies suggested that endogenous sources of new knowledge, together with a dynamism in entrepreneurial environment, institutions and cultures, are key factors for the capacity of the local economy to adapt and react to external shocks. Similarly, the absence of these features may contribute to the lack of resilience and the exacerbation of regional inequalities.

Martin and Gardiner (2019) investigated the resilience of 85 British cities to major economic shocks over the period 1971–2015. Cities showed a distinct shift in terms of resistance and recovery between the shocks and differences between northern and southern cities. The authors made an attempt to identify the main factors responsible for these patterns and, moreover, to estimate the potential impact of the Brexit shock. In Table 2.1, the authors synthesize the possible factors causing differences in the resilience to major shocks that may also contribute to the long-run growth paths of cities (Table 2.1).

Table 2.1 Possible influences in city economic resilience (Source: Martin and Gardiner 2019)

Type of influence	Possible effects
Economic structure and market orientation	The more diversified (more specialized) the city's economic structure, the more (less) resilient it will be to shocks The greater a city's dependence of manufacturing, the less resilient it will be to shocks, given this sector's traditionally higher elasticity of demand The greater a city's dependence on services, the more resilient it will be to shocks, given this sector's lower elasticity of demand The more localized (geographically dispersed) are city's industrial supply chains, the less (more) resilient it will be to shocks
The size of the public sector	Given the traditional stability of public services over the economic cycle, the greater a city's dependence on public sector the more resilient it will be to shocks Conversely, large-scale cuts in local public services associated with fiscal consolidation policies may reduce the "buffering" role of the public sector
Scale and nature of export base	The impact on a city's resilience will depend on the nature of its export base: It may shield a city from internal (domestic) shocks, but may expose it to demand or other shocks originating in its overseas export markets
Competitiveness of local firms	The more productive a city's firms, the more resilient they will be to shocks since they should have a competitive advantage over less productive cities A high innovation rate, which helps drive firm competitiveness, should make for enhanced firm adaptability and greater city resilience
Skill base and labour market flexibility	The presence of high-skilled, well-qualified workers is widely thought to influence a city's economic dynamism; thus, to the extent that such workers are more productive, and their skills more transferable or adaptive, the more resilient a city should be to shocks Skilled workers are more likely to be able to move into the more dynamic parts of a city's economy and thus assist recovery from shocks A steady in-migration of skilled workers may thus improve a city's economic resilience, while, conversely, the sustained outflow of skilled workers could well erode a city's resilience
Size of city economy (agglomeration economies)	Large cities are claimed to benefit from various agglomeration economies that help make local firms more productive and innovative, and that attract high-skilled workers. Large cities also tend to be more economically diverse. Together these features should make larger cities more resilient to shocks
Firms' access to finance and credit	The availability and commitment of local sources of loan finance or capital, including low costs and favourable terms of credit, may help small local firms to weather downturns and maintain or re-orientate production and employment more easily than under conditions where finance is restricted
Governance and policy regime	A city with a well-organized, consensual and strategic economic governance structure committed to short- and long-run policies aimed at supporting businesses and jobs may improve a city's resistance to and/or its recoverability from shocks

Nevertheless, evidence on the determinants of regional resilience is rather scarce and when existing, it regards regions within a single country. Recent empirical analyses on regional economic resilience follow the evolutionary approach (Simmie and Martin 2010; Hassink 2010; Martin and Sunley 2015; Martin and Gardiner 2019; Boschma 2015; Giacometti and Teräs 2019; Doran and Fingleton 2018). The useful insights arising from these investigations is that, rather than simply recovering from a short-term shock, cities or regions are more successful in preventing a decline the more they are able to develop new growth paths by boosting new industries or technological changes. A diversified industrial mix characterized by low sectoral interdependencies, for example, may support a greater regional resistance. Other key factors are the following: skilled and innovative workforce, modern productive infrastructures, highly developed knowledge networks (universities-local industries-firms), supportive financial systems, liberal market conditions and policy activism (Christopherson et al. 2010; Desrochers and Leppälä 2011; Martin 2012; Boschma 2015; Di Caro 2015; Sensier et al. 2016). Moreover, a full investigation of resilience requires to consider also the governance and the leadership to understand what makes a region more resilient than others (Brooks et al. 2016). Public sector exerts a key role as resilience can be favoured by a clear leadership, strategic approaches, high public sector skills and a transparent government (OECD 2016).

2.4 Regional Resilience to Climate-Related Emergencies and Disasters

A less salient but a progressively more disruptive source of unsettling shocks for regions is climate change, whose prominence in the public debate has risen—ironically—while the economic systems were struggling with the adverse consequences of the "double-dip" recession. As a matter of fact, macroeconomic policy and climate policy are crossing paths with growing frequency, as demonstrated by proposals of "green" monetary policy. Fostering regional resilience may eventually involve some coordination among policy tools that so far have belonged to separate domains.

Yet, if the focus is shifted to the developing world, regional resilience to climate change is decisively conditioned by trade, agricultural and development policy programmes. Indeed, among the major factors increasing pressures on such vulnerable areas, climate change and environmental degradation occupy prominent positions, as increasingly shown in the econometric literature (Carleton and Hsiang 2016; Beine and Parsons 2015; Gray and Mueller 2012a, b; Mastrorillo et al. 2016), threatening agricultural productivity and causing losses in the already thin infrastructural endowments in rural poor areas (Barrios et al. 2006; Marchiori et al. 2012; Piguet et al. 2011; Bozzola et al. 2018). The literatures dealing with the impact of disasters (Altay and Ramirez 2010; Caschili et al. 2015; Porcelli and Trezzi 2019) and of disease spread (Wu et al. 2016; Costello et al. 2009; Semenza and Menne 2009) add further layers

Table 2.2 Alternative approaches to measuring regional economic resilience (based on Martin and Sunley 2015)

Method/focus	Approach
Descriptive case studies	
Mainly narrative-based, using simple descriptive data, interviews with key actors, interrogation of policies, etc. may be comparative (e.g. two cities or regions)	Simmie and Martin (2010, Cambridge and Swansea) Treado and Wolfe (2010; Pittsburgh) Wolfe (2010; Ottawa and Waterloo) Hill et al. (2012, Seattle) Cowell (2013, Cleveland) Enelow (2013, Detroit) Evans and Karecha (2014, Munich) Hu and Hassink (2017, Chinese regions)
Resilience indices	
Singular or composite measures, often relative to some reference position May involve "dashboards" using key economic indicators Often comparative (several cities, regions), to produce "resilience rankings" of cities and regions	Martin (2012, UK regions) Augustine et al. (2013) Han and Goetz (2015, US counties) Rockefeller Foundation (2015, World cities) Martin et al. (2016, UK regions) Salvati et al. (2017, Italian Cities) Angulo et al. (2018, Spanish regions) Ibl, Svoboda, and Siegert (2018, EU regions) Sensier (2018, EU regions) Spencer (2018, US cities)
Statistical time-series models	
ARIMA-type models (with dummies for shock and recovery periods) to generate counterfactual or expected position of city or region assuming no shock, against which actual outcomes are compared Also stochastic mean reversion models	Fingleton et al., (2012, UK regions) D'Lima and Medda (2015, London)
Causal and structural models	
Used either to estimate counterfactual positions or include dummies for shocks among regressors Used to generate impulse responses or error correction measures Includes models that regress resilience indices on selected "causal" variables	Doran and Fingleton (2013, US Cities) Fingleton and Palombi (2013, British towns) Fingleton et al. (2015, EU regions) Salvati et al. (2017, Italian regions) Pudelko and Hundt (2017, German regions) Kitsos and Bishop (2018, Britain's local authority districts) Sprague (2018, US counties)

of knowledge on the overall impact of climate change, as the associated shocks are partly and perhaps increasingly due to the changing global climate.

The threat of low resilience to climate shocks in developing countries can easily be grasped upon reading Table 2.2 sourced from Martin and Gardiner (2019). Lack of sectoral diversification, high dependence on externally provided services, highly localized supply chains, low quality of public sector services, backward technologies and worker skills, lack of agglomeration economies, highly inefficient credit markets and short-term oriented economic governance—these are all negative influences on resilience that hamper the prospects for long-term economic growth in developing countries.

In relation to climate change, resilience is commonly understood to embrace both the capacity to withstand the disruptive effects of climate change and the speed and flexibility of response and recovery. As such, regions can improve their resilience if appropriate mitigation and adaptation activities are put in place. More specifically, adaptation can increase the speed of recovery, whereas mitigation can make the system more robust to shocks (McDaniels et al. 2008).

Adaptation is defined by IPCC (Data Distribution Centre—Glossary) as "The process of adjustment to actual or expected climate and its effects" adding that "In human systems, adaptation seeks to moderate harm or exploit beneficial opportunities".[1] Instances of adaptation measures comprise precautionary measures (e.g. evacuation plans), investments in the security of buildings and infrastructures, landscape reforestation.

Instead, mitigation is defined as a "human intervention to reduce the sources or enhance the sinks of greenhouse gases (GHGs)" (IPCC Data Distribution Centre—Glossary). Examples of mitigation include the use of alternative transport means (car sharing, bicycles), an increasing penetration of renewable energy sources for electrical power and the adoption of energy-efficient technologies.

Adaptation is usually seen as a more flexible response to unexpected changes in climatic conditions, whereas mitigation may require structural change in production facilities, such as the de-commissioning of coal-fired power plants and investments in new renewable energy generation facilities.

Adaptation can be planned (e.g. policy-driven) or autonomous (Tanner and Mitchell 2008). It can be further classified into anticipatory and reactive (Tanner and Mitchell 2008; Cannon and Müller-Mahn 2010). Reactive adaptation refers to behaviours that individuals spontaneously enact in order to reduce their exposure to shocks. Anticipatory adaptation, instead, concerns all efforts made by individuals and policy-makers to improve the resilience of economic activities with respect to 000 possible future perturbations. Reactive adaptation has always been observed in history, whether the changing climate was due to anthropogenic factors or not.[2]

[1] A related concept is vulnerability that IPCC (Data Distribution Centre—Glossary) defines as "the propensity or predisposition to be adversely affected", encompassing "lack of capacity to cope or adapt". Turner et al. (2003) conceive vulnerability as a coupled human-environment system interaction associated with a likelihood of experiencing harm due to exposure to hazards.

[2] Adaptation can involve a response along an intensive margin or an extensive margin. An intensive margin response consists of an intensified or more frequent use of a certain behaviour or technology,

An interesting example of the built-in adaptation capacity of vulnerable areas is given by human migration. Historically, it has always represented an adaptive response to hardships caused by poverty and social deprivation, but more recently human migration has been conceived as a possible strategy to reduce the pressures on territories and populations affected by dramatic environmental and climatic changes.

However, the on-set of migration as an adaptive response is not granted and requires specific socio-economic conditions that may widely differ across and within regions. While climatic shocks deteriorate economic conditions and safety, fostering migration, they may also knock down households' wealth, thereby increasing the difficulties to migrate. The propensity to migrate, indeed, is not only a matter of mobility but also depends on the availability of social and financial resources necessary to move away (Kleemans 2015). People that are most exposed and vulnerable to climate change are not necessarily the ones most likely to migrate (Brown 2008).

The presence of mixed effects is confirmed by the empirical literature. Some studies find that outgoing migration flows from agriculturally dependent areas increase as temperatures hit crops, through either strong reduction in levels or moisture declines (Cattaneo and Peri 2016; Hornbeck 2012; Feng et al. 2010; Cai et al. 2016). Others, on the contrary, show that natural disasters influencing incomes, such as hurricanes and flooding, do not cause a concrete migration response in the poorest countries (Bohra-Mishra et al. 2014; Drabo and Mbaye 2011; Gray and Mueller 2012a, b).

In the seek for family's livelihoods, migrants look for better opportunities elsewhere. Before the whole family migrates from its own land, one or more family members are first sent away, in an attempt to diversify income, gain knowledge, spread risk and acquire capabilities to sustain the origin community (De Haan et al. 2000). This solution strengthens the family's capacity to offset idiosyncratic risks linked either to extreme weather events or other local specific characteristics, contributing to reduce the overall risk. In addition, migrant remittances may facilitate the rebuilding of houses and infrastructures or even to support a transformation process which favours adaptation. In this sense, while reinforcing the ability to take advantage of the new development trajectories driven by these changes, climate-induced migration may favour the concretization of the more encompassing notion of resilience. In the past, migration caused by climate changes was considered as a pathology to be prevented, especially because the unexpected and large-scale movements forced governments of hosting countries to face challenges that could overwhelm their management capacities and provoke conflict. Later it has been often considered as matter of refugee rights (Piguet 2013).

Recently, however, the shift towards resilience has reframed the debate. The possibility of sudden and unpredictable changes in the global system (Lenton et al. 2008) could in fact lead to an irreversible destruction of the Amazon rainforest or a breakdown of the Gulf Stream which could put at risk the containment of greenhouse gas concentrations in the atmosphere under safe thresholds. These challenges motivate the increasing efforts of international governmental and research institutions in cli-

whereas an extensive margin response occurs when a new technology or behaviour is added to the existing repertoire of adaptive responses.

mate science to make populations resilient to the effects of climate change. Among them, the Intergovernmental Panel on Climate Change (IPCC) adopted the concept of resilience for its 2012 special report. Hence, under this new global context, migration, which allows for the relocation of millions of people in different countries, becomes a rational and more accepted strategy of adaptation (Methmann and Oels 2015; Robalino et al. 2015; Scheffran et al. 2012). At the end, migrant societies may create new social capital and networks among themselves, connecting both host and home communities and empowering them to strengthen their resilience. This outcome is confirmed by one of the few empirical studies dealing with climate-change migration and resilience conducted by Scheffran et al. (2012) on a sample of Northwest African countries.

Developing countries rank among the most vulnerable also with respect to indirect effects of climate change, such as weather-related natural disasters. Indeed, international statistics on disasters prove that the developed world is the worst damaged by disasters in terms of casualties and infrastructural losses (Caschili et al. 2015), not less importantly because of limited endowments of competences and financial resources that would enable prompt and effective policy responses. Moreover, recent evidence reveals that the economic value of the damages correlated with climate change is increasing over time (Coronese et al. 2019). Academic efforts towards studying the shock propagation mechanisms and formulating effective policy responses have highlighted the need for network-based and regional economic paradigms (Caschili et al. 2015).

As a matter of fact, natural disasters can have pervasive effects across sectors and firms and through the supply chains connecting them. Yet, disasters are not all alike: the damages from windstorms and floods, which are more closely "climatic" disasters, seem to differ quite significantly from that of an earthquake. For this reason, an all-hazards approach is valuable (Altay and Ramirez 2010). This is even more true if one considers that, although earthquakes are not caused by climate change, adaptation by the oil and gas industries to mitigation strategies by competing energy sources (such as renewables) and the threat of resource exhaustion has led to the development of fracking technologies that are suspected of increasing the frequency of earthquakes. The extant literature on resilience to natural disasters, in any case, shows that the depth and length of the recovery period vary according to the position of the economy, sector or firms in the supply chain (Altay and Ramirez 2010) and depend on the offsetting role of reconstruction policies put in place in the aftermath of the event (see, e.g. Porcelli and Trezzi 2019 and references therein).

2.5 Resilience Measurement and Empirical Strategies

The measurement of economic resilience represents another relevant issue in the reference literature. There are various approaches to measuring resilience which consist on describing regions' own ability (adaptive capacity) to address a sudden shock or on measuring the outcomes of its efforts in reacting at the shock. However, a successful result in coping a negative shock does not necessarily imply that a region will be resilient also in the future (Sensier et al. 2016). Since a region may lack the capacity to build adequate efforts to overcome future shocks, the evaluation of its resilience should go beyond a simple measure of its performances, in terms of income, employment, trading potential and so on. It becomes crucial to identify the factors enhancing its capacity to resist, respond and recover from a shock.

Resilience is mainly measured with reference to macroeconomic variables such as employment, output, wages and international trade links, at varying degrees of territorial disaggregation like national, regional and cities (Sensier et al. 2016). From this indicative list, however, in case of a recessionary shock researchers prefer to use a measure of the employment change as it enables to capture also the social aftermaths of a given shock (Fratesi and Rodríguez-Pose 2016). The rationale is that, as employment takes longer to return to its pre-crisis level, greater efforts are required from policy-makers in order to tackle social problems following any severe economic shock (Reinhart and Rogoff 2009).

Martin (2012) suggests to measure regional resistance through a "sensitivity index" (β_r) that compares the percentage change in employment in a given region with the change at the national level following a negative downturn of the economy. Indicating with E the level of employment (E_r and E_N, respectively, at regional and national levels) and with ΔE the correspondent changes, the sensitivity index for any region is built as follows:

$$\beta_r = \frac{\Delta E_r / E_r}{\Delta E_N / E_N} \tag{2.1}$$

If β_r is greater than one, the region shows a low resistance to a recessionary shock as compared to the country, while values lower than one imply a high level of regional resistance (low sensitivity to shocks).

Generally, empirical studies (Martin 2010; Martin et al. 2016; Palaskas et al. 2015; Doran and Fingleton 2018, among others) are focused on two elements of regional resilience: resistance, which is the ability of a region to resist the initial impact of the crisis and recovery, the ability to recover after the crisis. The suggestion is to measure both resistance and recoverability of a region with respect to an expected position, which could be either where the region would have been, had a shock not occurred or the observed change at national level after the shock (Martin and Gardiner 2019). In the first case, the pre-shock growth path is projected forward using statistical time-series or appropriate structural models in order to predict the benchmark position. Logically, this implies a strong assumption, namely, that everything should remain unchanged regarding structure and features of the region itself.

The second procedure, aimed at comparing the capacity of reaction of a region with another, implies a comparison between the regional change and the national one taken as a benchmark. The indicators for resistance and recoverability are, respectively,

$$
\begin{aligned}
Resis_i &= \frac{\Delta E_i^c - \Delta \hat{E}_i^c}{E_i^0} \\
Recov_i &= \frac{\Delta E_i^r - \Delta \hat{E}_i^r}{E_i^0}
\end{aligned}
\tag{2.2}
$$

where ΔE_i^c is the actual change in employment in region i during the contraction period or the economic crisis and ΔE_i^r is the actual change in employment after the crisis during the recovery period. Counterfactual changes (based on national changes) are $\Delta \hat{E}_i^c$ and $\Delta \hat{E}_i^c$, respectively, for the contraction and the recovery periods. Difference between actual and benchmark values is scaled by the initial level of employment in region i (E_i^0). Both measures of resilience are centred on zero, in which case region performs as the macro-aggregation, in the sense that it resists or recovers as the national economy does. Conversely, negative values indicate a relative weak (lower) resilience of the region and positive values reveal a stronger relative resilience (resistance/recovery). This approach, differently from the first one, does not take into account the specific structure of each region, nor its geographical position which may be crucial in shaping regional performance.

Resilience strongly depends on the specific context of a region and is place-based, and hence the choice of the indicator requires knowledge of the local preconditions. The OECD (2014) recommends to look at the different kinds of capital which may influence regional resilience (financial, human, natural, physical, political and social) to understand the risky factors for a given economy and build appropriate policy responses. An overview of the methodologies more frequently employed in the empirical analyses may be found in Martin and Sunley (2015), Doran and Fingleton (2016).

These methodologies range from descriptive analyses, that usually employ case studies comparing geographical units on the basis of more or less complex indexes (Simmie and Martin 2010; Evans and Karecha 2014; Bailey and Berkeley 2014), to more sophisticated statistical and econometric ones. Case studies have the advantage of following a place-based perspective that favours, more than others, the choice of the better indicator to evaluate a region's resilience. Statistical and econometric approaches are usually employed to explore the extent to which pre-existing conditions shape the ability of a region to resist to or to recover from a shock. Econometric analyses may follow either a time-series approach, more robust but requiring long time periods (Fingleton et al. 2012), or panel data approaches, based on multidimensional measurements of a number of local units over time.

If viewed from the perspective of climate economics, the above resilience measures suffer from being narrowly designed to cope with the methodological challenges posed by business cycle frequency fluctuations. This is all the more evident when attempting to select a credible counterfactual and measure resistance to and recoverability from climate change. Econometric work incorporating climate expectations does exist, e.g. Severen et al. (2018) on land value, but the informative value of such measures is questionable in light of the complex and highly non-linear dynamics characterizing climate change. In general, benchmarks based on time-series models or causal-structural models, as summarized in Table 2.2, are doomed to misfire since they are fed by retrospective information. But nothing guarantees that econometric coefficients and the distribution of the error term will remain stable, nor that econometric models have sufficient external validity as to inform policy-makers on potential damages in the future.

Not by chance, agencies studying the threat of climate change, such as IPCC, rely on numerical methodologies for prediction and assessment, such as Integrated Assessment Models (IAMs). Such models manage to overcome the restrictive assumptions of econometric models, such as (log)-linearity and spherical disturbances, and are not bound by the availability of historical data.[3] Though the resilience measures discussed in this section prove valuable when attempting to assess the effects of climate change on a shorter time horizon, and on markets and sectors driven by relatively high frequency dynamics. Two among the sectors that are most exposed to the impacts of climate change, agriculture and energy, will be the subjects of Chaps. 5 and 6 in this book, respectively. In both cases, the data frequency is sufficiently high as to warrant the construction of counterfactuals based on econometric projections in space or time. This is true for the resilience of rural areas as measured through balance-sheet data concerning land value, which is available at annual frequency, and even more for electricity markets that publish data and prices and trading volumes every hour.

2.6 A Prime Overview of Spatial Dependence in Regional Economic Resilience: The Added Value of Spatial Analysis

The empirical analyses on regional resilience focused mainly on the search for the characteristics that would make regions more resilient and to assist policy-makers in formulating appropriate measures to reduce the vulnerability of spatial systems to shocks and enhance their ability to better respond to and recover from the crises (Bristow and Healy 2018; Crespo et al. 2014; Wink 2014).

However, a strand of theoretical analyses sheds light on the presence of spatial interactions, that is, the rise of beneficial or harmful externalities generated by policies implemented in neighbouring regions (Kelejian and Robinson 1993; Solé-Ollé 2006). The reasons why this occurs are various, from intergovernmental competition (Buettner 2001) to the case of policy "mimicking", when voters judge the competence of their own politicians comparing their performances with the neighbours' ones (Salmon 1987; Besley and Case 1995). To account for the presence of externalities and spillover mechanisms, empirical analyses may relate to spatial econometric approaches that model the statistical dependence due to geographical as well as socio-economic proximity (Anselin 1988; Corrado and Fingleton 2012).

---

[3] For more information on these modelling approaches and on the possibly fruitful combination of econometric and simulative models, please refer to Chap. 5, Sect. 5.5 in this book.

Consideration of these mechanisms becomes necessary, since failure to acknowledge the presence of spatial effects would result in model mis-specification (LeSage and Pace 2009), due to serious problems of omitted variables, and lead to an incorrect representation and understanding of the true causal processes at work. The main consequences may be a bias in estimation results and destabilizing policy indications (Hammadou et al. 2014; Baicker 2005; Fowles and Tandberg 2017).

So far, spatial econometrics modelling in regional economic resilience analysis has been used to obtain counterfactual predictions on income or employment conditions (wages and employment levels) to be compared with the actual paths that followed a negative economic shock. The main aim of this type of comparisons is to investigate the transmission of economic shocks across different geographical locations and to test whether the responsiveness of each of them may be influenced by the presence of spatial interdependencies.

Recently, empirical studies started to follow different methodologies, as, for example, spatial panel models, accounting for the presence of spatial interdependencies within places (Fingleton and Palombi 2013), or procedures merging individual data with regional data (Doran and Fingleton 2016, 2018).

The first study analysing local economies resilience to recessionary shocks in a context of spatial interdependence is that of Fingleton and Palombi (2013). In particular, their study analyses the relative resilience of UK towns to the major recessionary shocks occurred in the historical period 1871–1906. By combining the insights of the literature on resilience and spatial econometrics methods to model the transmission of the shocks, they show that highly and increasingly specialized towns were relatively more prone to shocks. The sectoral composition of employment, therefore, turns out to be crucial for the economic resilience. The more the productive activity is diversified, the higher the ability of a town to adapt to a shock and even to outperform their counterfactual paths by the end of the post-shock period.

Given the acknowledgment on the validity of spatial panel data model for forecasting purposes, Angulo et al. (2018) followed Fingleton and Palombi (2013) in order to evaluate the impact of the 2007 crisis on the annual employment growth rate of the Spanish provinces. Different spatial panel specifications were initially used to analyse the evolution of the employment before the crisis, from 1980 to 2006. After choosing the best model, the authors used the estimation results to forecast employment annual growth rates over the period after the crisis and considered them the counterfactual values. Finally, counterfactual values have been compared to actual ones in order to measure the responsiveness of each province to the economic crisis. Using a localization quotient as a measure of economic specialization, the authors explored also the possibility that the crisis did not equally affect all sectors. In doing so, they found that sectors such as non-market services and construction showed a higher vulnerability to the economic shock while specialization on energy, manufacturing, transport and common services reinforced the resilience of provinces helping them to return to their pre-shock growth paths.

Another study that confirms the importance of the industrial structure on the economic resilience is the analysis of the impact of the 2008 crisis on the US metropolitan areas by Doran and Fingleton (2018). The main contributions of this paper are the use of a spatial dynamic panel model, to account for dynamic as well as for spatial interactions, and the introduction of covariates that enable to obtain consistent results in presence of problems of omitted variables and potential endogeneity of regressors. The key findings of this analysis are, first, that highly specialized metropolitan areas appear to be more severely affected by the crisis and, second, that during the post-crisis period a structural change may help the recovery of the area.

Differently from other studies focusing on aggregated data at city, region or country levels, Doran and Fingleton (2016) use micro-level data for a sample of 13 European countries to analyse individual employment resilience to the 2008 economic crises. More precisely, this paper follows an empirical methodology that merges individual and regional data. Data from the European Social Survey (ESS) over the period 2002–2008 are taken to investigate the influence of individual characteristics such as education, age and regional unemployment rate on individual employment resilience by controlling also for potential labour market spillover effects across regions. Results reveal that Central Europe regions appear to be more resilient than peripheral ones, with regions in Ireland, Spain and Portugal most severely affected by the 2008 crisis. As regards individual characteristics, high-skilled individuals are more resilient than the low skilled, suggesting that education increases the possibility to find a job, and in addition it makes employees more resilient to negative economic shocks. Moreover, mid-aged individuals are more resilient than younger and older individuals.

References

Altay, N., & Ramirez, A. (2010). Impact of disasters on firms in different sectors: implications for supply chains. *Journal of Supply Chain Management, 46*(4), 59–80.

Angulo, A. M., Mur, J., & Trivez, F. J. (2018). Measuring resilience to economic shocks: An application to Spain. *Annals of Regional Science, 60*, 349–373.

Anselin, L. (1988). *Spatial conometrics: Methods and models*. Dordrecht, The Netherlands: Kluwer Academic Publishers.

Augustine, N., Wolman, H., Wial, H., & McMillen, M. (2013). *Regional economic capacity, economic shocks and economic resilience*. MacArthur Foundation Network on Building Resilient Regions Working Paper, May.

Baicker, K. (2005). The spillover effects of state spending. *Journal of Public Economics, 89*, 529–544.

Bailey, D., & Berkeley, N. (2014). Regional responses to recession: The role of the West Midlands regional taskforce. *Regional Studies, 48*(11), 1797–1812.

Barrios, S., Bertinelli, L., & Strobl, E. (2006). Climatic change and rural-urban migration: The case of sub-Saharan Africa (No. 06-01). FEDEA.

Beine, M., & Parsons, C. (2015). Climatic factors as determinants of international migration. *The Scandinavian Journal of Economics, 117*(2), 723–767.

Besley, T., & Case, A. (1995). Incumbent behavior: Vote seeking, tax setting and yardstick competition. *American Economic Review, 85*(1), 25–45.

Blanchard, O. T., & Katz, L. F. (1992). Regional evolutions. *Brookings Papers on Economic Activity, 1*, 1–75.

Bohra-Mishra, P., Oppenheimer, M., & Hsiang, S. M. (2014). Nonlinear permanent migration response to climatic variations but minimal response to disasters. *Proceedings of the National Academy of Sciences, 111*(27), 9780–9785.

Bonß, W. (2016). The notion of resilience: Trajectories and social science perspectives. In A. Mauer (Ed.), *New perspectives on resilience in socio-economic spheres* (pp. 9–24). E-book: Springer VS.

Boschma, R. (2015). Towards an evolutionary perspective on regional resilience. *Regional Studies, 49*, 733–751.

Bozzola, M., Massetti, E., Mendelsohn, R., & Capitanio, F. (2018). A Ricardian analysis of the impact of climate change on Italian agriculture. *European Review of Agricultural Economics, 45*(1), 57–79.

Brand, F., Hoheisel, D., & Kirchhoff, T. (2011). Der Resilienz-Ansatz auf dem Prüfstand: Herausforderungen, Probleme, Perspektiven. In Bayerische Akademie für Naturschutz und Landschaftspflege (Hrsg.), Landschaftsökologie. Grundlagen, Methoden, Anwendungen (pp. 78–83). Laufen: ANL.

Bristow, G. (2010). Resilient regions: Re-placeing regional competitiveness. *Cambridge Journal of Regions, Economy and Society, 3*(1), 153–167.

Bristow, G., & Healy, A. (Eds.) (2018). *Economic crisis and the resilience of regions: A European study*. Cheltenham: Edward Elgar.

Brooks, C., Vorley, T., & Williams, N. (2016). The role of civic leadership in fostering economic resilience in city regions. *Policy Studies, 37*(1), 1–16.

Brown, O. (2008). Migration and climate change (No. 31). United Nations Publications.

Buettner, T. (2001). Local business taxation and competition for capital: The choice of the tax rate. *Regional Science and Urban Economics, 31*(2–3), 215–245.

Cambridge Journal of Regions, Economy and Society (2010). *The Resilient Region, 3*(1), 1–167.

Caballero, R., & Hammour, M. (1994). The cleansing effect of recession. *American Economic Review, 84*, 1350–1368.

Cai, R., Feng, S., Oppenheimer, M., & Pytlikova, M. (2016). Climate variability and international migration: The importance of the agricultural linkage. *Journal of Environmental Economics and Management, 79*, 135–151.

Cannon, T., & Müller-Mahn, D. (2010). Vulnerability, resilience and development discourses in context of climate change. *Natural Hazards, 55*(3), 621–635.

Carleton, T. A., & Hsiang, S. M. (2016). Social and economic impacts of climate. *Science, 506*, 3.

Caschili, S., Reggiani, A., & Medda, F. (2015). Resilience and vulnerability of spatial economic networks. *Networks and Spatial Economics, 15*(2), 205–210.

Cattaneo, C., & Peri, G. (2016). The migration response to increasing temperatures. *Journal of Development Economics, 122*, 127–146.

Cerra, V., & Saxena, S. C. (2008). Growth dynamics: The myth of economic recovery. *American Economic Review, 98*(1), 439–457.

Christopherson, S., Michie, J., & Tyler, P. (2010). Regional resilience: Theoretical and empirical perspectives. *Cambridge Journal of Regions, Economy and Society, 3*(1), 3–10.

Coronese, M., Lamperti, F., Keller, K., Chiaromonte, F., & Roventini, A. (2019). Evidence for sharp increase in the economic damages of extreme natural disasters. *Proceedings of the National Academy of Sciences, 116*(43), 21450–21455.

Corrado, L., & Fingleton, B. (2012). Where is the economics in spatial econometrics? *Journal of Regional Science, 52*, 210–239.

Costello, A., Abbas, M., Allen, A., Ball, S., Bell, S., Bellamy, R., et al. (2009). Managing the health effects of climate change: Lancet and University College London Institute for Global Health Commission. *The Lancet, 373*(9676), 1693–1733.

Cowell, M. (2013). Bounce back or move on: Regional resilience and economic development planning. *Cities, 30*, 212–222.

Crespo, J., Suire, R., & Vicente, J. (2014). Lock-in or lock-out? How structural properties of knowledge networks affect regional resilience. *Journal of Economic Geography, 14*(1), 199–219.
De Haan, A., Brock, K., Carswell, G., Coulibaly, N., Seba, H., & Toufique, K. A. (2000). *Migration and livelihoods: Case studies in Bangladesh*. Institute of Development Studies, Brighton, Sussex: Ethiopia and Mali.
Desrochers, P., & Leppälä, S. (2011). Creative cities and regions: The case for local economic diversity. *Creativity and Innovation Management, 20*(1), 59–69.
Di Caro, P. (2015). Recessions, recoveries and regional resilience: Evidence on Italy. *Cambridge Journal of Regions, Economy and Society, 8*(2), 273–291.
Di Caro, P., & Fratesi, U. (2018). Regional determinants of economic resilience. *Annals of Regional Science*, *60*, 235–240.
D'Lima, M., & Medda, F. (2015). A new measure of resilience: An application to the London underground. *Transportation Research: Policy and Practice, 81*, 35–46.
Doran, J., & Fingleton, B. (2013). US metropolitan area resilience: Insights from dynamic spatial panel estimation. *Environment and Planning A: Economy and Space, 50*, 111–132.
Doran, J., & Fingleton, B. (2016). Employment resilience in Europe and the 2008 economic crisis: Insights from micro-level data. *Regional Studies*, *50*, 644–656.
Doran, J., & Fingleton, B. (2018). US metropolitan area resilience: Insights from dynamic spatial panel estimation. *Environment and Planning A: Economy and Space*, *50*(1), 111–132.
Dosi, G., & Nelson, R. R. (1994). An introduction to evolutionary theories in economics. *Journal of Evolutionary Economics*, *4*(3), 153–172.
Drabo, A., & Mbaye, L. M. (2011). Climate change, natural disasters and migration: An empirical analysis in developing countries (IZA Discussion Papers 5927). Bonn: Institute for the Study of Labor (IZA).
Enelow, N. (2013). The resilience of Detroit: An application of the adaptive cycle metaphor to an American metropolis, Paper, Economics for Equity and Environment.
Eraydin, A. (2012). Resilient thinking for planning. In A. Eraydin, & T. Taşan-Kok (Eds.), *Resilient Thinking in Urban Planning* (pp. 17–37). Dordrecht, Heidelberg, New York & London: Springer.
Evans, R., & Karecha, J. (2014). Staying on top: Why is Munich so resilient and successful? *European Planning Studies, 22*, 1259–1279.
Feng, S., Krueger, A. B., & Oppenheimer, M. (2010). Linkages among climate change, crop yields and Mexico-US cross-border migration. *Proceedings of the National Academy of Sciences U.S.A.*, *107*, 14257–14262.
Fingleton, B., Garretsen, H., & Martin, R. L. (2012). Recessionary shocks and regional employment. *Journal of Regional Science*, *52*(1), 109–133.
Fingleton, B., Garretsen, H., & Martin, R. L. (2015). Shocking aspects of monetary union: The vulnerability of regions in Euroland. *Journal of Economic Geography*, *15*, 907–934.
Fingleton, B., & Palombi, S. (2013). Spatial panel data estimation, counterfactual predictions, and local economic resilience among British towns in the Victorian era. *Regional Science and Urban Economics*, *43*(4), 649–660.
Fowles, J. T., & Tandberg, D. A. (2017). State higher education spending: A spatial econometric perspective. *American Behavioral Scientist, 61*(14), 1773–1798.
Fratesi, U., & Rodríguez-Pose, A. (2016). The crisis and regional employment in Europe: What role for sheltered economies? *Cambridge Journal of Regions, Economy and Society, 9*(1), 33–57.
Friedman, M. (1993). The plucking model of business fluctuations revisited. *Economic Enquiry*, *31*, 171–177.
Gabriel, T. (2005). Resilienz - Kritik und Perspektiven. *Zeitschrift für Pädagogik*, *51*(2), 208–218.
Gali, J., & Hammour, J. L. (1993). *Long run effects of business cycles*. New York: Columbia-Graduate School of Business.
Giacometti, A., & Teräs, J. (2019). *Regional economic and social resilience: An exploratory in-depth study in the Nordic countries*, Nordregio Report 2019, 2.

Gray, C. L., & Mueller, V. (2012a). Natural disasters and population mobility in Bangladesh. *Proceedings of the National Academy of Sciences, 109*(16), 6000–6005.

Gray, C. L., & Mueller, V. (2012b). Drought and population mobility in rural Ethiopia. *World Development, 40*(1), 134–145.

Gunderson, L., & Holling, C. S. (Eds.). (2002). *Panarchy: understanding transformations in human and natural systems* (pp. 25–62). Washington: Island Press.

Gunderson, L., Allen, C., & Holling, C. S. (Eds.). (2010). *Foundations of ecological resilience*. Washington: Island Press.

Günther, E. (2009). *Klimawandel und Resilience Management. Interdisziplinäre Konzeption eines entscheidungsorientierten Ansatzes*. Wiesbaden: Gabler.

Hammadou, H., Paty, S., & Savona, M. (2014). Strategic interactions in public R&D across European countries: A spatial econometric analysis. *Research policy, 43*(7), 1217–1226.

Han, Y., & Goetz, S. J. (2015). The economic resilience of US counties during the great recession. *The Review of Regional Studies, 45*, 131–149.

Hassink, R. (2010). Regional resilience: A promising concept to explain differences in regional economic adaptability. *Cambridge Journal of Regions, Economy and Society, 3*(1), 45–58.

Hill, E., St Clair, T., Wial, H., Wolman, H., Atkins, P., Blumenthal, P., & Friedhoff, A. (2012). Economic shocks and regional economic resilience. In *Building resilient regions* (pp. 193–274). New York University: Wagner Institute.

Holling, C. S. (1973). Resilience and stability of ecological systems. *Annual Review of Ecology and Systematics, 4*, 1–23.

Hornbeck, R. (2012). The enduring impact of the American dust bowl: Short and long-run adjustments to environmental catastrophe. *American Economic Review, 102*, 1477–1507

Hudson, R. (2010). Resilient regions in an uncertain world: Wishful thinking or a practical necessity? *Cambridge Journal of Regions, Economy and Society, 3*, 11–26.

Hu, X., & Hassink, R. (2017). Exploring adaptation and adaptability in uneven economic resilience: A tale of two Chinese mining regions. *Cambridge Journal of Regions, Economy and Society, 10*, 527–541.

Kelejian, H., & Robinson, D. (1993). A suggested method of estimation for spatial interdependent models with autocorrelated errors and an application to a country expenditure model. *Papers in Regional Science, 72*(3), 297–312.

Kleemans, M. (2015). Migration choice under risk and liquidity constraints (No. 330-2016-13981).

Lenton, T. M., Held, H., Kriegler, E., Hall, J. W., Lucht, W., Rahmstorf, S., et al. (2008). Tipping elements in the Earth's climate system. *Proceedings of the National Academy of Sciences, 105*(6), 1786–1793.

LeSage, J. P., & Pace, R. K. (2009). *Introduction to spatial econometrics*. New York: CRC Press.

Marchiori, L., Maystadt, J. F., & Schumacher, I. (2012). The impact of weather anomalies on migration in sub-Saharan Africa. *Journal of Environmental Economics and Management, 63*(3), 355–374.

Martin, R. L. (2010). The Roepke lecture in economic geography - rethinking regional path dependence: Beyond lock-in to evolution. *Economic Geography, 86*, 1–27.

Martin, R. (2012). Regional economic resilience, hysteresis and recessionary shocks. *Journal of Economic Geography, 12*(1), 1–32.

Martin, R. L. (2018). Shocking aspects of regional development: Towards an economic geography of resilience. In G. L. Clark, M. Feldman, M. Gertler, & D. Wojcik (Eds.), *The new Oxford handbook of economic geography* (pp. 837–864). Oxford: Oxford University Press.

Martin, R., & Gardiner, B. (2019). The resilience of cities to economic shocks: A tale of four recessions (and the challenge of Brexit). *Papers in Regional Science, 98*, 1801–1832.

Martin, R. L., & Sunley,. (2015). On the notion of regional economic resilience: Conceptualisation and explanation. *Journal of Economic Geography, 14*, 1–42.

Martin, R., Sunley, P., Gardiner, B., & Tyler, P. (2016). How regions react to recessions: Resilience and the role of economic structure. *Regional Studies, 50*(4), 561–585.

Mastrorillo, M., Licker, R., Bohra-Mishra, P., Fagiolo, G., Estes, L. D., & Oppenheimer, M. (2016). The influence of climate variability on internal migration flows in South Africa. *Global Environmental Change, 39*, 155–169.

McDaniels, T., Chang, S., Cole, D., Mikawoz, J., & Longstaff, H. (2008). Fostering resilience to extreme events within infrastructure systems: Characterizing decision contexts for mitigation and adaptation. *Global Environmental Change, 18*(2), 310–318.

Mergenthaler, A. (2012). *Gesundheitliche Resilienz. Konzept und Empirie zur Reduzierung gesundheitlicher Ungleichheit im Alter*. Wiesbaden: Springer VS.

Methmann, C., & Oels, A. (2015). From 'fearing' to 'empowering' climate refugees: Governing climate-induced migration in the name of resilience. *Security Dialogue, 46*(1), 51–68.

Modica, M., & Reggiani, A. (2015). Spatial economic resilience: Overview and perspectives. *Networks and Spatial Economics, 2*, 211–233.

Muštra, V., Šimundi, B., & Kuliš, Z. (2016). Effects of smart specialization on regional economic resilience in EU. In *First SMARTER Conference on Smart Specialisation and Territorial Development: Changing Patterns of Territorial Policy: Smart Specialisation & Innovation in Europe*.

OECD (2014). *Guidelines for resilience systems analysis*. OECD Publishing.

OECD (2016). Resilient cities, policy highlights of the OECD report (Preliminary version). The Organization for Economic Co-operation and Development(OECD).

Palaskas, T., Psycharis, Y., Rovolis, A., & Stoforos, C. (2015). The asymmetrical impact of the economic crisis on unemployment and welfare in Greek urban economies. *Journal of Economic Geography, 15*(5), 973–1007.

Pendall, R., Foster, K. A., & Cowell, M. (2010). Resilience and regions: Building understanding of the metaphor. *Cambridge Journal of Regions, Economy and Society, 3*(1), 71–84.

Porcelli, F., & Trezzi, R. (2019). The impact of earthquakes on economic activity: Evidence from Italy. *Empirical Economics, 56*(4), 1167–1206.

Piguet, E. (2013). From "primitive migration" to "climate refugees": The curious fate of the natural environment in migration studies. *Annals of the Association of American Geographers, 103*(1), 148–162.

Piguet, E., Pecoud, A., & de Guchteneire, P. (2011). *Migration and climate change*. Cambridge: Cambridge University Press.

Pike, A., Dawley, S., & Tomaney, J. (2010). Resilience, adaptation and adaptability. *Cambridge Journal of Regions, Economy and Society, 3*, 59–70.

Ramlogan, R., & Metcalfe, J. S. (2006). Restless capitalism: a complexity perspective on modern capitalist economies. In *Complexity and co-evolution: Continuity and change in socio-economic systems* (pp. 115–146).

Reinhart, C. M., & Rogoff, K. S. (2009). The aftermath of financial crises. *American Economic Review, 99*(2), 466–472.

Robalino, J., Jimenez, J., & Chacon, A. (2015). The effect of hydro-meteorological emergencies on internal migration. *World Development, 67*, 438–448.

Rockefeller Foundation (2015). *City resilience framework—100 resilient cities*. New York City and London: Rockefeller Foundation and Arup.

Rodin, J. (2015). *The resilience dividend: Managing disruption, avoiding disaster and growing strong in an unpredictable world*. London: ProfiGe Books.

Salmon, P. G. (1987). Decentralization as an incentive scheme. *Oxford Review of Economic Policy, 3*(2), 24–43.

Salvati, L., Carlucci, M., & Venanzoni, G. (2017). Recession, resilience, local labour markets: Wealthier is better? *Letters in Spatial and Resource Sciences, 10*, 177–204.

Scheffran, J., Marmer, E., & Sow, P. (2012). Migration as a contribution to resilience and innovation in climate adaptation: Social networks and co-development in Northwest Africa. *Applied Geography, 33*, 119–127.

Semenza, J. C., & Menne, B. (2009). Climate change and infectious diseases in Europe. *The Lancet Infectious Diseases, 9*(6), 365–375.

Sensier, M., Bristow, C., & Healey, A. (2016). Measuring regional economic resilience across Europe: Operationalising a complex concept. *Spatial Economic Analysis, 11*, 128–151.
Setterfield, M. (2010). *Hysteresis* (p. 1004). Economics, WP No: Trinity College of Dublin.
Severen, C., Costello, C., & Deschenes, O. (2018). A forward-looking Ricardian approach: Do land markets capitalize climate change forecasts? *Journal of Environmental Economics and Management, 89*, 235–254.
Simmie, J., & Martin, R. (2010). The economic resilience of regions: Towards an evolutionary approach. *Cambridge Journal of Regions, Economy and Society, 3*(1), 27–43.
Solé-Ollé, A. (2006). The effects of party competition on budget outcomes: Empirical evidence from local governments in Spain. *Public Choice, 126*(1–2), 145–176.
Tanner, T., Mitchell, T. (2008). Entrenchment or enhancement: could climate change adaptation help reduce poverty? Chronic Poverty Research Centre Working Paper (106).
Treado, C., & Wolfe, D. A. (2010). The strategic management of core cities: Path dependence and economic adjustment in resilient regions. *Cambridge Journal of Regions, Economy and Society, 3*, 139–152.
Turner, B. L., Kasperson, R. E., Matson, P. A., McCarthy, J. J., Corell, R. W., Christensen, L., et al. (2003). A framework for vulnerability analysis in sustainability science. *Proceedings of the National Academy of Sciences, 100*(14), 8074–8079.
Walker, B., Holling, C. S., Carpenter, S. R., Kinzig, A. (2004). Resilience, adaptability and transformability in socialecological systems. *Ecology and Society, 9*(2).
Webber, D. J., Healy, A., & Bristow, C. (2018). Regional growth paths and resilience: A European analysis. *Economic Geography, 94*, 355–375.
Werner, E. E. (1977). *The children of Kauai. A longitudinal study from the prenatal period to age ten*. Honolulu: University of Hawaii Press.
Werner, E. E., & Smith, R. S. (1982). *Vulnerable but invincible: A longitudinal study of resilient children and youth*. New York: McGraw Hill.
Werner, E. E., & Smith, R. S. (2001). *Journeys from childhood to midlife: Risk, resilience and recovery*. Ithaca, NY: Cornell University Press.
Wink, R. (2014). Regional economic resilience: Policy experiences and issues in Europe. *Raumforschung und Raumordnung, 72*(2), 83–84.
Wolfe, D. A. (2010). Pittsburgh's evolving steel legacy and the steel technology cluster. *Cambridge Journal of Regions, Economy and Society, 3*, 105–120.
Wu, X., Lu, Y., Zhou, S., Chen, L., & Xu, B. (2016). Impact of climate change on human infectious diseases: Empirical evidence and human adaptation. *Environment International, 86*, 14–23.
Zolli, A., & Healy, A. M. (2012). *Resilience: Why things bounce back*. New York: Simon and Shuster.

Chapter 3
Spatial Econometric Models: Theory

Abstract Accounting for the presence of spatial linkages is extremely important in regional economics. These linkages may follow, for example, when policies implemented by a specific geographical unit spread also to its proximate locations. In this regard, spatial analysis provides different tools to detect the presence of spatial interdependencies. The failure to acknowledge for these externalities would result in a mis-specification of the empirical model, with an incorrect representation and understanding of the true causal processes at work. Readers interested in understanding the way this transmission mechanism takes place, and how different regions react to external shocks, would highly benefit from learning the basics of spatial econometrics. To this aim, the chapter starts from a brief description of the Exploratory Spatial Data Analysis (ESDA), providing global and local indexes to control for the presence of spatial patterns, verifying whether there is a clustering in the space of high or low outcomes for the regional variable under observation. The chapter then illustrates the main models of spatial interaction: depending on the type of interaction between observations of neighbouring units, different specifications may be followed (SAR, SEM, SDM, SARAR and so forth) according to the results of model selection tests (usually Likelihood-based tests).

Keywords Spatial dependence · Spatial weight matrix · Exploratory spatial data analysis · Model estimation · Cross-sectional data · Panel data · Spatial econometrics

3.1 Introduction to Spatial Analysis

The growing attention on the locational attributes of phenomena observed in the natural and social environments has strengthened the commitment of scholars to deal with spatial information in their theoretical and empirical frameworks. The relevance of space has been summarized by Tobler (1970) in the First Law of Geography, stating that "Everything is related to everything else, but near things are more related than distant things". A crucial aspect becomes the identification of the surrounding locations that may exercise an influence on a given data point. Indeed, if spatial

R. De Siano et al., *Regional Resilience to Climate and Environmental Shocks*, SpringerBriefs in Regional Science, https://doi.org/10.1007/978-3-030-54588-8_3

interdependence among data points does exist, more sophisticated models are needed in order to estimate relationships influenced by spatial linkages, going beyond non-spatial regressions.

In this view, regional resilience assessments may benefit from testing whether employment, income or other relevant variables in one region may be related to what happens in neighbouring regions. Spatial linkages, indeed, may affect first of all the mechanism responsible for the propagation of a shock across regions and, second, they contribute to spread the outcomes of local policies across geographical boundaries. It follows that both beneficial and harmful externalities spill over the most proximate regions, and ignoring these linkages may lead to an incorrect representation and understanding of the true causal processes at work. There is a large evidence that, when proper corrections for spatial effects are not included in the model specification, econometric analysis of spatially distributed variables can lead to incorrect statistical inference (Voss et al. 2006).

Readers interested in understanding the way such transmission mechanisms take place, and how different regions react to external shocks, would highly benefit from learning the basics of spatial econometrics. To this aim, the present chapter starts from a brief description of the Exploratory Spatial Data Analysis (ESDA), which provides both global and local indexes to check whether spatial patterns do exist; in other words, whether there is a clustering in the space of high or low outcomes for the regional variable/characteristic under observation.[1] The chapter then illustrates the main models of spatial interaction (Anselin 1988a; LeSage and Pace 2009; Elhorst 2010, 2011): depending on the type of interaction between observations of neighbouring units, and according to the results of model selection tests (usually Likelihood-based tests), different specifications may be adopted (SAR, SEM, SDM, SARAR and so forth).

For the sake of completeness, the spatial interaction analysis may also concern flows between locations (points, areas) or between nodes of a network. This is the case with flows between an origin and a destination point/area in a geographic space, usually related to transport, population migration, commerce, tourism, journey-to-work and transmission of information and knowledge across space. Spatial interaction models typically used to represent these flows are of the gravity type, as they are aimed at detecting the ability of origins to generate flows, the attractiveness of destinations and the factors that may prevent the interaction (Fischer 2010). These models are usually employed in frameworks different from that of spatial resilience. Hence, in this book, we have preferred to focus our overview on models and tools that are more appropriate in analysing regional resilience. For an exhaustive discussion of spatial interaction modelling, we refer the interested reader to Fischer and Wang (2011) and Patuelli and Arbia (2016) among others.

[1] Further details on spatial descriptive statistics are available in Chap. 4 and its Appendix.

3.2 Modelling Space

Modelling spatial dependence requires an appropriate representation of relative spatial positions. When considering a sample of R spatial units, the solution is given by a spatial neighbours matrix (**W**), that is, a square symmetric R by R matrix with the (i, j) element equal to 1 if spatial units i and j are neighbours of one another (or, more generally, are spatially related), and zero otherwise. By convention, the diagonal elements of this spatial neighbours matrix are set to zero. In the simplest form, the element w_{ij} represents the spatial proximity based on the concept of binary contiguity. If two spatial units have a common border they will be considered contiguous and will be marked with value 1. Conversely, if they are not contiguous, their coupling will have a value of 0.

$$\mathbf{W} = \begin{bmatrix} 0\,1\,1\,0 \\ 1\,0\,1\,0 \\ 1\,1\,0\,1 \\ 0\,0\,1\,0 \end{bmatrix}$$

However, the simple contiguity measure has some limits: (a) it does not account for non-reciprocal interactions because of its symmetrical nature; (b) it does not account for other kind of interaction than geographical proximity; (c) it does not distinguish between different types of neighbouring regions, with regard to the distance or the morphology of the border area (mountain, hill, plain) or, finally, to the length of the border actually shared. To overcome the latter problem contiguity matrices of higher order than the first can be used.[2] Besides, it is possible to associate the contiguity matrix with a matrix of distances or with other matrices that combine distance measurements with border lengths. In this line, one can use the inverse of the distance between the centres (geographical, but also political or administrative) of spatial units and the distance can either be expressed in terms of linear and road distance or in terms of time distance, in relation to travel times. The matrix obtained, in this way, assigns to each pair of spatial units a different value that decreases with the distance to account for the stronger influence that closer spatial unit may exert with each other.

$$\mathbf{W} = \begin{bmatrix} 0 & \frac{1}{d_{1,2}} & \frac{1}{d_{1,3}} & \frac{1}{d_{1,4}} \\ \frac{1}{d_{2,1}} & 0 & \frac{1}{d_{2,3}} & \frac{1}{d_{2,4}} \\ \frac{1}{d_{3,1}} & \frac{1}{d_{3,2}} & 0 & \frac{1}{d_{3,4}} \\ \frac{1}{d_{4,1}} & \frac{1}{d_{4,2}} & \frac{1}{d_{4,3}} & 0 \end{bmatrix}$$

[2]For example, a second-order matrix identifies those areas that are contiguous to regions showing the value one in the first-order matrix.

The first and second limits are usually overcome by introducing measures of "economic contiguity" between spatial units, like commercial exchanges (which would also distinguish non-reciprocal relationships) or the degree of similarity in productive specialization. Finally, proximity may account also for other types of distances such as cultural, linguistic or administrative.

The best choice regarding spatial neighbours matrix is made taking into account the research goals, that is, ensuring that measures of proximity included in the matrix are strongly exogenous with respect to the variable object of study.

The spatial neighbours matrices are usually transformed into spatial weight matrices following the most common approach of the row-standardization, consisting in dividing each row element by the sum of all the elements in the row:

$$w_{ij}^{*} = \frac{w_{ij}}{\sum w_{ij}}$$

Row-standardized spatial weight matrices are particularly attractive because their elements can be interpreted as the fraction of total spatial influence on unit i attributable to unit j and, when weights are based on the inverse distances, they decrease with increasing distances.

3.3 Exploratory Spatial Data Analysis

Following Anselin (1996), Exploratory Spatial Data Analysis (ESDA) represents a useful set of instruments to test for the presence of spatial dependence. ESDA is a collection of techniques aimed to describe and visualize spatial distributions; identify atypical locations or spatial outliers; discover patterns of spatial association, clusters or hotspots and suggest spatial regimes or other forms of spatial heterogeneity. Central to this conceptualization is the notion of spatial autocorrelation or spatial association, i.e. the phenomenon where locational similarity (observations in spatial proximity) is matched by value similarity (attribute correlation). If either high or low values are found in a spatial unit and its adjacent, it would be an instance of positive spatial autocorrelation. Spatial clustering is said to occur with positive spatial autocorrelation (Anselin et al. 2000). By contrast, negative spatial autocorrelation would occur if units with low levels for the variable of interest surround a spatial unit with a high value. For those spatial units where no correlation exists between variables values and their locations, the spatial pattern is considered to exhibit zero spatial autocorrelation (Holt 2007).

The literature distinguishes between global and local spatial autocorrelation tests. Global spatial autocorrelation tests aim to test for the presence of potential clusters of high/low outcome levels across regions. Local spatial autocorrelation tests go beyond locating the clusters, measuring their spatial extent and identifying the regions which contribute more to global autocorrelation.

The most common test for the existence of global spatial autocorrelation is due to Patrick Moran and is usually referred to as Moran's I statistic (1948):

$$I = \frac{\sum_{i=1}^{R} \sum_{j=1}^{R} w_{ij}(x_i - \mu)(x_j - \mu)}{\sum_{i=1}^{R} (x_i - \mu)^2} \tag{3.1}$$

where x_i represents the variable describing the phenomenon under study in region i, μ is the sample mean and w_{ij} is the weight of a row-standardized spatial matrix. The expected value of the Moran index ($E(I)$) is equal to $-1/(R-1)$. With R large enough the standardized I is normally distributed, and therefore the rejection of the null hypothesis of no spatial autocorrelation implies the presence of spatial dependence. Values of I greater than the expected value indicate positive spatial autocorrelation, which means that regions with high (low) values tend to be located close to other regions with high (low) levels. Values of I less than the expected value indicate a negative association, and hence a tendency for dissimilar values in nearby regions.

Local indicators of spatial clustering analysis consider the relationship between each region and its neighbours, identifying hotspots (high-value clusters) and cold spots (low-value clusters). Among these tests there are the Getis-Ord statistic (Getis and Ord 1992, 1995; Sokal et al. 1998), the Moran scatterplot (Anselin 1996) and the Local Indicator of Spatial Association (LISA) (Anselin 1995).

The Getis-Ord test refers to the concentration of values of the variable of interest in the neighbourhood of region i. The original statistic is as follows:

$$G_i = \frac{\sum_{j=1}^{R} w_{ij} x_j}{\sum_{j=1}^{R} x_j} \tag{3.2}$$

with $j \neq i$, where w_{ij} is the corresponding element of a non-standardized symmetric binary weights matrix which attributes 1 to neighbouring regions and 0 to the others and to the pivot region. Once standardized, positive values of G_i indicate spatial clustering of highly values around region i, while negative values indicate a cluster of regions showing low values.

The Moran scatterplot is given by a graph centred on the mean value of the variable of interest x for a number of spatial units (see Fig. 3.1 for an example). The horizontal axis returns the standardized values of the variable while the vertical one returns the spatially lagged value of the variable ($\mathbf{W}x$) (Anselin 2005; Anselin et al. 2000). The slope of the linear regression line that runs through the scatter plot is the Moran's-I coefficient (Anselin et al. 2000). If the points are equally dispersed between the four quadrants this will indicate no correlation (the slope is zero). If, however, there is a clear relationship, the Moran scatter plot can be used to distinguish different types of spatial autocorrelation depending on the quadrant. High-high (upper right) and low-low (lower left) represent positive spatial autocorrelation (spatial clusters) and low-high and high-low are negative spatial autocorrelation (spatial outliers) (Anselin 2005; Anselin et al. 2000).

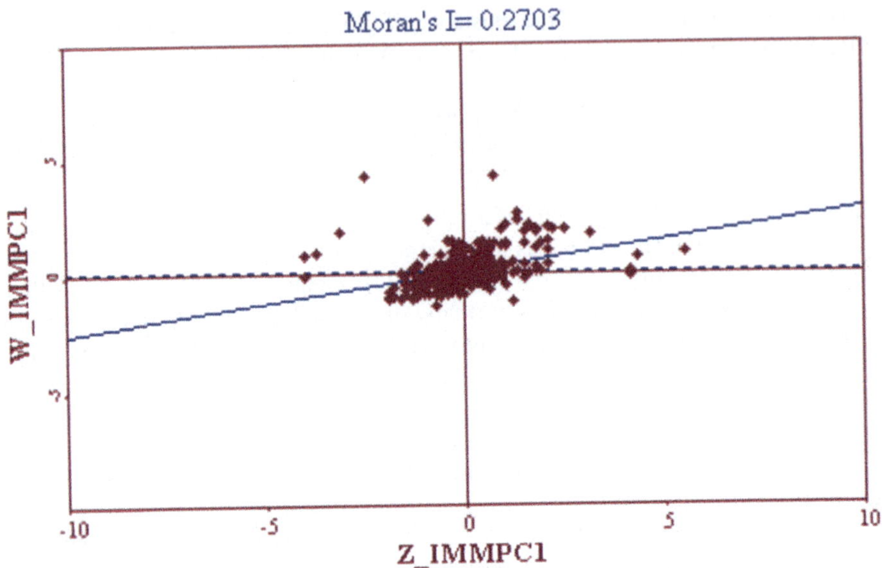

Fig. 3.1 Moran scatter plot: an example. *Source* authors' own elaboration

Since Moran scatter plots do not assess the statistical significance of spatial associations it is useful to associate this global indicator, and its graphical representation, with a local autocorrelation indicator that is able to measure interdependence for each of the regions concerned. The Local Indicator of Spatial Association (LISA) effectively enables for each spatial unit to associate a measure of the spatial association level with its surroundings and to assess its statistical significance. The indicator commonly used as LISA is the local version of Moran's I statistic for each spatial unit i:

$$I_i = \frac{x_i - \mu}{m_0} \sum_{j=1}^{R} w_{ij}(x_j - \mu) \tag{3.3}$$

with $m_0 = \sum_{j=1}^{R} \frac{(x_i-\mu)^2}{R}$. A positive value for I_i indicates spatial clustering of similar values (high or low) whereas a negative value indicates spatial clustering of dissimilar values between region i and its neighbours.

3.4 Spatial Econometrics Models

The spatial econometric literature distinguishes three generations of spatial models, respectively, for cross-sectional data and static or dynamic panel data approaches. There are also different ways of modelling spatial dependence (Anselin 1988a), depending on the type of interaction between observations of neighbouring units observed. This section aims at presenting the most common model specifications employed for empirical spatial analyses in the different data contexts.

3.4.1 Cross-Section Models

In a cross-section data context models are formalized as follows:

- Spatial Autoregressive Model (SAR), when levels of the dependent variable y depend on the levels of y in the neighbouring regions (spatial lag dependence) and on a set of observed local characteristics. In this case, the formal model is

$$\mathbf{y} = \lambda \mathbf{W}\mathbf{y} + \mathbf{X}\beta + \epsilon \tag{3.4}$$

  where λ is the spatial autoregressive parameter, which has to be estimated from the data, $\mathbf{W}$ is a predefined R by R spatial weighting matrix, $\mathbf{X}$ is a vector of explanatory variables, measuring other regional characteristics and ϵ is an independently and identically distributed error term for region i with zero mean and variance σ^2. $\mathbf{Wy}$ represents the spatially lagged values of the dependent variable accounting for the idea of spatial spillover within nearby regions. The significance of λ indicates the presence of spatial autocorrelation, that is, the extent of interaction between the observations according to the spatial pattern exogenously introduced with the standardized weight matrix $\mathbf{W}$.
  For an individual observation, the basic spatial lagged autoregressive equation is simply:

$$y_i = \lambda \sum_j w_{ij} y_j + X_i \beta + \epsilon_i \tag{3.5}$$

- Spatial Error Model (SEM), when error terms are correlated across space (spatial error dependence). This situation may occur when some determinants of the dependent variable omitted from the model are spatially autocorrelated or when unobserved shocks follow a spatial pattern. The resulting model is written as follows:

$$\begin{aligned} \mathbf{y} &= \mathbf{X}\beta + \epsilon \\ \epsilon &= \rho \mathbf{W}\epsilon + \mathbf{v} \end{aligned} \tag{3.6}$$

  with $\mathbf{v}$ assumed to be normal with zero mean and variance $\sigma^2\mathbf{I}$.

- Spatial Autocorrelation Model (SAC) that combines the SAR and the SEM models assuming that spatial dependence runs through both the dependent variable and the error term.[3] The model is as follows:

$$\begin{aligned} \mathbf{y} &= \lambda \mathbf{W}_1 \mathbf{y} + \mathbf{X}\beta + \epsilon \\ \epsilon &= \rho \mathbf{W}_2 \epsilon + \mathbf{v} \end{aligned} \tag{3.7}$$

where $\mathbf{W}_1$ and $\mathbf{W}_2$ may or may not be equal to each other.
- Spatial Durbin Model (SDM) containing a spatially lagged dependent variable and spatially lagged independent variables.

$$\mathbf{y} = \lambda \mathbf{W}\mathbf{y} + \mathbf{X}\beta + \mathbf{W}\mathbf{X}\theta + \epsilon \tag{3.8}$$

The spatially lagged dependent variable included in the SDM accounts for the endogenous interaction effects emerging when the choice of country i is influenced by decisions taken in neighbouring countries j (with $i \neq j$), while the spatially lagged independent variables capture the exogenous interaction effects arising from explanatory variables. The meaning is that changes in the explanatory variables occurring in a region may directly affect the decision taken in that region (direct effect) and, in addition, the decisions taken in surrounding regions which, in turn, exert an influence back to the region itself (indirect effects). This influence corresponds to the spatial spillovers emerging between proximate geographical units. The total impact of any explanatory variable on a single region's outcome results from a combination of both its direct and indirect (neighbourhood) effects.

In this regard, the spatial Durbin regression model is particularly useful since it computes summary measures of these varying impacts separately (LeSage and Pace 2014; LeSage and Dominguez 2012). In precise terms, the partial derivative of the dependent variable with respect to the explanatory variable (given by the average of the main diagonal elements of the n by n matrix of partial derivatives) is taken as a measure of the average direct effect. The average of off-diagonal elements of the n by n matrix of partial derivatives, instead, is a measure of the indirect or spillover effect. The sum of these two measures provides a summary measure of the total impact associated with changes in explanatory variables on the dependent variable. Another reason enhancing the use of the SDM is that it produces unbiased estimates even if the data-generation process is any other spatial model specification (Greene 2005).

[3]The model is alternatively referred to as the spatial autoregressive with Spatially Autocorrelated Errors (SARAR).

3.4.2 Panel Data Models

Recently, the spatial econometrics analysis has shifted to model specifications based on panel datasets containing time-series observations for all the spatial units under observation (Anselin et al. 2008; Elhorst 2010, 2011). Indeed, when dealing with geo-referenced data it is important to account also for space-specific and time-invariant variables using spatial fixed effects that may capture determinants difficult to measure which lead to bias in the estimates. Similarly, there are time-specific and space-invariant factors that can be accounted for by using time fixed effects. The latter enable to capture events such as economic expansion or recessions, changes in government policies or environmental shocks and so on.

In what follows, a brief overview of spatial panel data version of the models is presented for the cross-section data context:

- Spatial Autoregressive Model (SAR), in the case of a spatial lag dependence, the model is

$$\mathbf{y}_t = \lambda \mathbf{W} \mathbf{y}_t + \mathbf{X}_t \beta + \mu + \epsilon_t \tag{3.9}$$

  where λ is the spatial autoregressive parameter, $\mathbf{W}$ the spatial weights matrix and, for each period $t = 1, \ldots, T$, $\mathbf{y}_t$ is the $n \times 1$ column vector of the dependent variable, $\mathbf{X}_t$ the $n \times k$ matrix of regressors, μ is the vector of spatial-specific effects and ϵ_t the vector of error terms with the standard assumption $\epsilon_{it} \sim N(0, \sigma^2)$ and $E(\epsilon_{it}\epsilon_{jt}) = 0$ for $i \neq j$ and $t \neq s$. Spatial-specific effect may be treated as fixed or as random, depending on the outcome of Hausman's specification test (Baltagi 2008). The null hypothesis of the test is of random effects, and therefore, if it is rejected, the random effects model is rejected in favour of the fixed effects one.
- Spatial error model, based on spatial autocorrelation in the error term:

$$\begin{aligned} \mathbf{y}_t &= \mathbf{X}_t \beta + \mu + \epsilon_t \\ \epsilon_t &= \lambda \mathbf{M} \epsilon_t + \mathbf{v}_t \end{aligned} \tag{3.10}$$

- Spatial Durbin model, representing a generalization of the SAR model which also includes spatially weighted independent variables as explanatory variables:

$$\mathbf{y}_t = \lambda \mathbf{W} \mathbf{y}_t + \mathbf{X}_t \beta + \mathbf{W} \mathbf{Z}_t \theta + \mu + \epsilon_t \tag{3.11}$$

  A generalization of the model may be obtained using different spatial weights for the spatially lagged dependent variable ($\mathbf{Wy}$) and the spatially weighted regressors ($\mathbf{WZ}$) or by using $\mathbf{Z}_t \neq \mathbf{X}_t$.
- Spatial autocorrelation model, combining the SAR with a spatial autoregressive error model:

$$\mathbf{y}_t = \lambda \mathbf{W} \mathbf{y}_t + \mathbf{X}_t \beta + \mu + \epsilon_t \quad (3.12)$$

$$\epsilon_t = \lambda \mathbf{M} \epsilon_t + \mathbf{v}_t \quad (3.13)$$

where **M** is a matrix of spatial weights which may or may not be equal to **W**.

The specifications of spatial panel models considered up to now are static, in the sense that they consider contemporaneous values of the dependent and independent variables. The spatial econometric literature, however, presents also dynamic specifications. In what follows, we present the dynamic variant of the SAR models:

$$\mathbf{y}_t = \tau \mathbf{y}_{t-1} + \psi \mathbf{W} \mathbf{y}_{t-1} + \lambda \mathbf{W} \mathbf{y}_t + \mathbf{X}_t \beta + \mu + \epsilon_t \quad (3.14)$$

where two components are added to the SAR specification, one for the dependent variable lagged in time and another for the dependent variable lagged both in space and in time. Similarly, can be done for the SDM specification, adding an additional component for the regressors lagged both in time and in space.

3.4.3 *Specification Tests*

Since the spatial dependence structure between countries may be very complicated, the regression model chosen for cross-national/regional empirical analysis has to be able to exploit the wealth of information these relationships contain.

The spatial autocorrelation tests presented in Sect. 3.2 are powerful in revealing the presence of spatial dependence but do not allow to discriminate between the different types of spatial interactions (Anselin et al. 1996). To this purpose, the existence of decision rules based on spatial autocorrelation tests helps to choose between the multiplicity of specifications. To identify the proper spatial econometric model specification (Anselin et al. 2010; Elhorst 2010, 2011; LeSage and Pace 2009), one may perform the two Lagrange multiplier tests, LMLAG and LMERR (Anselin 1988b), and their robust counterparts, RLMLAG and RLMERR (Anselin et al. 1996), using the baseline empirical model without any spatial component. The null hypothesis of these tests is the absence of spatial dependence: LMLAG for an autoregressive spatial lag variable and LMERR for a spatial autocorrelation of errors. The two robust tests RLMLAG and RLMERR have good power against their specific alternatives. The decision rule suggested by Anselin and Florax (2012) enables to decide which specification is the more appropriate. If LMLAG is more significant than LMERR and RLMLAG is significant but RLMERR is not, then the appropriate model is the spatial autoregressive model. Conversely, if LMERR is more significant than LMLAG and RLMERR is significant but RLMLAG is not, then the appropriate specification is the spatial error model.[4]

[4]This procedure is known as from General to Specific.

Besides, following LeSage and Pace (2009); Elhorst (2010, 2011) one can decide to start with the more general SDM specification and then test for alternative spatial dependence models by using specific log-likelihood ratio tests indicating whether the SDM may collapse alternatively to the Spatial Autoregressive (SAR), the Spatial Error (SEM) or the Spatial Autocorrelation (SAC) model. Test results and AIC (Akaike 1974) and BIC (Schwarz 1978) information criteria will help the researcher to identify the appropriate model specification.[5]

3.4.4 Estimation Approaches

Regarding the estimation procedures, it is important to remember that the spatially lagged dependent variable (**Wy**) is correlated with the errors, even when these are identical and independently distributed. In other words, the spatial lag of the dependent variable should always be considered as endogenous. This means that the OLS estimation approach becomes inadequate because producing biased and inconsistent estimators (typically those deriving from simultaneity problems). Among the alternative methodologies, chosen to overcome these drawbacks, the most widely used is the Maximum Likelihood (ML) estimation technique (Ord 1975). Indeed, under specific conditions,[6] all the classic properties of its estimator (non-distortion, efficiency and asymptotic normality) remain valid when including spatially lagged dependent variables.

Other estimation approaches are the following: quasi-maximum likelihood, QML (Lee 1925); instrumental variables, IV (Anselin 1988a); generalized method of moments, GMM (Kelejian and Prucha 1998, 1999) and Bayesian Markov chain Monte Carlo methods, MCMC (LeSage 1997). In particular, Fingleton and Le Gallo (2007, 2008), Drukker et al. (2013), Liu and Lee (2013) show that IV/GMM estimators are extremely useful in those cases where linear spatial dependence models contain one or more endogenous explanatory variables (other than the spatially lagged dependent variable) that need to be instrumented.

References

Akaike, H. (1974). A new look at the statistical model identification. IEEE Transactions on Automatic Control AC-19, 716–723.

Anselin, L. (1988a). *Spatial Econometrics: Methods and Models*. Dordrecht, The Netherlands: Kluwer Academic Publishers.

Anselin, L. (1988b). Lagrange multiplier test diagnostics for spatial dependence and spatial heterogeneity. *Geographical Analysis*, *20*, 1–17.

[5]The rule is to choose the spatial model with the lowest information criteria value.

[6]The conditions regard derivability of the function, the existence of partial derivatives and the existence of a positive and non-singular covariance matrix (Bates and White 1985).

Anselin, L. (1995). Local indicators of spatial association - LISA. *Geographical Analysis, 27*, 93–115.
Anselin, L. (1996) The Moran scatterplot as an ESDA tool to assess local instability in spatial association. In M. M. Fisher, H. J. Scholten & D. Unwin (Eds.), *Spatial analytical perspectives on GIS*. London, UK: Taylor & Francis.
Anselin, L. (2005). Exploring spatial data with GeoDa: A workbook. http://geodacenter.asu.edu/system/files/geodaworkbook.pdf.
Anselin, L., & Florax, R. (Eds.). (2012). *New directions in spatial econometrics*. Springer Science & Business Media.
Anselin, L., Bera, A. K., Florax, R., & Yoon, M. J. (1996). Simple diagnostic tests for spatial dependence. *Regional Science and Urban Economics, 26*, 77–104.
Anselin, L., Cohen, J., Cook, D., Gorr, W., & Tita, G. (2000). Spatial analysis of crime. *Criminal Justice, 4*, 213–262.
Anselin, L., Le Gallo, J., & Javet, H. (2008). Spatial panel Econometrics. In L. Matyas & P. Sevestre (Eds.), *The econometrics of panel data, fundamentals and recent developments in the theory and practice* (pp. 625–660). Dordrecht: Kluwer.
Anselin, L., Syabri, I., & Kho, Y. (2010). GeoDa: an introduction to spatial data analysis. In *Handbook of applied spatial analysis* (pp. 73–89). Berlin, Heidelberg: Springer.
Baltagi, B. (2008). *Econometric analysis of panel data*. New York: Wiley.
Bates, C., & White, H. (1985). A unified theory of consistent estimation for parametric models. *Econometric Theory, 1*(2), 151–178.
Drukker, D. M., Egger, P., & Prucha, I. R. (2013). On two-step estimation of a spatial autoregressive model with autoregressive disturbances and endogenous regressors. *Econometric Reviews, 32*, 686–733.
Elhorst, J. P. (2010). Spatial panel data models. In M. M. Fischer, & A. Getis (Eds.), *Handbook of applied spatial analysis*. Berlin: Springer.
Elhorst, J. P. (2011). Matlab software for spatial panels. http://www.regroningen.nl/elhorst/doc/Matlab-paper.pdf
Fingleton, B., & Le Gallo, J. (2007). Finite sample properties of estimators of spatial models with autoregressive, or moving average disturbances and system feedback. *Annales d'economie et de statistique, 87*(88), 39–62.
Fingleton, B., & Le Gallo, J. (2008). Estimating spatial models with endogenous variables, a spatial lag and spatially dependent disturbances: Finite sample properties. *Papers in Regional Science, 87*, 319–339.
Fischer, M. M. (2010). Spatial interaction models. In B. Wharf (Ed.), *Encyclopedia of geography* (pp. 2645–2647). London: Sage.
Fischer, M. M., & Wang, J. (2011). *Spatial data analysis: models, methods and techniques*. Springer Briefs in Regional Sciences.
Getis, A., & Ord, J. K. (1992). The analysis of spatial association by use of distance statistics. *Geographical Analysis, 24*, 189–206.
Getis, A., & Ord, J. K. (1995). Local spatial autocorrelation statistics: Distributional issues and an application. *Geographical Analysis, 27*(4), 286–306.
Greene, W. H. (2005). *Econometric analysis* (6th ed.). Upper Saddle River, NJ: Pearson Prentice Hall.
Holt, J. B. (2007). The topography of poverty in the United States: A spatial analysis using county-level data from the community health status indicators project. *Preventing Chronic Disease, 4*(4), 1–9.
Kelejian, H. H., & Prucha, I. R. (1998). A generalized spatial two stage least squares procedure for estimating a spatial autoregressive model with autoregressive disturbances. *Journal of Real Estate Finance and Economics, 17*(99), 121.
Kelejian, H. H., & Prucha, I. R. (1999). A generalized moments estimator for the autoregressive parameter in a spatial model. *International Economic Review, 40*(509), 533.

Lee, L.-F. (2004). Asymptotic distribution of quasi-maximum likelihood estimators for spatial autoregressive models. *Econometrica, 72*, 1899–1925.
LeSage, J. P. (1997). Bayesian estimation of spatial autoregressive models. *International Regional Science Review, 20*, 113–129.
LeSage, J. P., & Dominguez, M. (2012). The importance of modeling spatial spillovers in public choice analysis. *Public Choice, 150*(3–4), 525–545.
LeSage, J. P., & Pace, R. K. (2009). *Introduction to spatial econometrics*. New York: CRC Press.
LeSage, J. P., & Pace, R. K. (2014). Interpreting spatial econometric models. *Handbook of regional science* (pp. 1535–1552). Berlin: Springer.
Liu, X., & Lee, L. F. (2013). Two stage least squares estimation of spatial autoregressive models with endogenous regressors and many instruments. *Econometric Reviews, 32*, 734–753.
Moran, P. (1948). The interpretation of statistical maps. *Journal of the Royal Statistical Society, Series B, 10*, 243–51.
Ord, K. (1975). Estimation methods for models of spatial interaction. *Journal of the American Statistical Association, 70*(349), 120–126.
Patuelli, R., & Arbia, G. (Eds.). (2016). *Spatial econometric interaction modelling*. Springer International Publishing.
Schwarz, G. (1978). Estimating the dimension of a model. *The Annals of Statistics, 6*, 461–464.
Sokal, R. R., Oden, N. L., & Thomson, B. A. (1998). Local spatial autocorrelation in a biological model. *Geographical Analysis, 30*, 331–54.
Tobler, W. (1970). A computer movie simulating urban growth in the detroit region. *Economic Geography, 46*(2), 234–240.
Voss, P. R., Long, D. D., & Hammer, R. B. (2006). County child poverty rates in the U.S.: A spatial regression approach. *Population Research Policy Review, 25*, 369–391.

Chapter 4
A Tutorial on Modelling Geographic, Economic and Social Interactions Using GIS Methods with R

Abstract Aim of this chapter is to briefly introduce practical techniques to work with economic data with a spatial component. In particular, it introduces a number of tools for modelling geographic and social/economic interactions, and it discusses most popular statistics to test for the presence of spatial autocorrelation. The reader is walked through the discussion with a number of replicable examples with R. Far from being a comprehensive survey of all methods existing in the literature, the dissertation is meant to provide a basic toolbox of investigation for the researcher which can be used and further extended in future studies.

Keywords GIS · R · Network analysis · Spatial statistics

4.1 Introduction

Aim of this chapter is to briefly introduce the reader to practical techniques for modelling geographic and social/economic linkages among spatial units using Geographical Information Systems (GIS).[1] The use of GIS is widespread in the economic literature. It is often employed to map economic data with a spatial component and generate additional spatial data as inputs to statistical analysis.[2] Moreover, GIS can introduce economics to new data complementing official statistics, when these are poorly measured or considered unreliable. This is the case when measuring economic growth (Henderson et al. 2012), the quality of political institutions (Hodler and Raschky 2014), the spread of illegal activities (Burgess et al. 2012) and the geo-

[1] The reader not acquainted with these methods is referred to Clarke (2003), Church and Murray (2009), Longley et al. (2005) for details.
[2] For a general introduction to GIS applications to economics see Overman (2008).

Electronic supplementary material The online version of this chapter (https://doi.org/10.1007/978-3-030-54588-8_4) contains supplementary material, which is available to authorized users.

R. De Siano et al., *Regional Resilience to Climate and Environmental Shocks*, SpringerBriefs in Regional Science, https://doi.org/10.1007/978-3-030-54588-8_4

graphical distribution of ethnicities into countries (Nunn 2008). In addition, GIS has been applied to the study of countries' development as determined by infrastructure networks (Dell 2015; Donaldson and Hornbeck 2016; Storeygard 2016), environmental policies (Chen et al. 2013) and conflicts (Michalopoulos and Papaioannou 2016; Rogall 2015; Yanagizawa-Drott 2014). Finally, GIS has been used to design credible identification strategies (Dell 2010; Dinkelman 2011; Duflo and Pande 2007; Durante et al. 2019; Michalopoulos and Papaioannou 2013; Olken 2009; Qian 2008).

With respect to standard spatial units of analysis (e.g. countries, administrative districts, etc.), a prominent feature of GIS-produced data is the ability to represent new artificial units of analysis—e.g. administrative units across countries, planned portions of infrastructure systems, locations of ethnic/linguistic groups—and explore different granularity levels at which economic patterns may prevail (see Cheshire and Sheppard 1995; Baum-Snow 2007; Michaels 2008; Michalopoulos and Papaioannou 2013, among others).

When spatial units are defined and their location has been identified, one can investigate whether and how spatial relations among units affect their economic outcomes. An inherent difficulty in this analysis is that geographic effects are often mediated by social interactions. The economic consequences of social network structures have been largely discussed in the literature (see Jackson and Zenou 2013; Jackson et al. 2017 for recent reviews), and it should not be surprising that they have been found to affect and be affected by spatial interactions.[3] On the one hand, it is well known that social connections decrease perceived physical distance, and this affects the formation of migration and trade flows (Abbate et al. 2012; Aten 1997; Koskinen and Lomi 2013), the diffusion of political information (Baybeck and Huckfeldt 2002) and the determinants of collaborations among organizations (Bevc et al. 2009; Lomi and Pallotti 2012) and scientists (Chandra et al. 2007). On the other hand, physical factors have an impact in the formation of social ties, for instance, friendship ties (Festinger et al. 1950; McPherson et al. 2001; Mouw and Entwisle 2006; Preciado et al. 2012; Warnes 1986).

This chapter introduces to software packages allowing to study such interplay between spatial and social/economic relations. To this purpose, the reader is walked through the discussion using replicable examples with R. Far from being a comprehensive survey of all methods existing in the literature, the dissertation is meant to provide a basic toolbox of investigation for the researcher which can be used and further extended in future studies.

[3]The analysis of social networks has a longstanding tradition. The interested may refer to Wasserman and Faust (1994) for a general introduction to this field of studies and to Jackson (2010) for applications to economics.

4.2 Chapter Overview

The exercise presented in this chapter is relative to the study of agro-food import flows among Sub-Saharan countries. There are three elements necessary to perform this exercise. The first is a map of Sub-Saharan countries. This map is stored in a "shapefile", that is, the GIS file format. The second element is the value of top 5 import flows among Sub-Saharan countries from 1990 to 2010. The third and last element is an adjacency matrix registering top 5 import flows among Sub-Saharan countries, where the generic cell i, j takes 1 if j is a top 5 importer of i, and 0 otherwise. Both the second and the third objects are constructed using the Food and Agriculture Organization Corporate Statistical Database (FAOSTAT, http://www.fao.org/faostat/en/#home).

In the first part of the exercise, the reader will learn how to load the data in the R environment, and it will be shown how to associate a coordinate system to the shapefile in order to localize countries and measure the distance between them. In the second part, it is discussed how spatial interactions can be measured. To this purpose, a number of routines will be introduced to register different spatial connections among countries. These will be displayed in the form of an adjacency matrix, where the ijth cell registers the presence and the strength of interaction between i and j. In the third part, it is presented how to work with social network data, and how to combine social network and spatial data. In the fourth and final part, the reader is introduced to the most popular statistics of spatial autocorrelation, which allow to test the extent to which a spatial unit is similar to other nearby units, and to identify the potential presence of externalities that may occur among units.

4.3 Load Data and Software Packages

In this section are presented the scripts to load the data and the software packages needed to replicate the exercise.

First, we load the software packages.

```
-----------------------------------------------------------
RUN COMMAND 1
-----------------------------------------------------------
```

Then, we download a shapefile containing the map of all world countries, and we load it into the R environment.

```
-----------------------------------------------------------
RUN COMMAND 2
-----------------------------------------------------------
```

Later on, we will need to merge the shapefile data with other data sources. For consistency with the format of the countries' name contained in the other data sources, we assign a lower capital name and a unique ID to each country contained in the shapefile.

Fig. 4.1 Map of the Sub-Saharan countries

```
--------------------------------------------------------------
RUN COMMAND 3
--------------------------------------------------------------
```

We create a data frame containing the name and the ID of Sub-Saharan countries.

```
--------------------------------------------------------------
RUN COMMAND 4
--------------------------------------------------------------
```

Finally, we use the object created with Command (4) to select in the shapefile the map of the Sub-Saharan countries, which can be visualized in Fig. 4.1.

```
--------------------------------------------------------------
RUN COMMAND 5
--------------------------------------------------------------
```

We now create the second and third elements necessary for the exercise. We begin with constructing an adjacency matrix where country i is connected to country j if j represents a top 5 importers of country i.

```
--------------------------------------------------------------
RUN COMMAND 6
--------------------------------------------------------------
```

Then we create a vector registering the value of top 5 import flows in each Sub-Saharan country.

```
--------------------------------------------------------------
RUN COMMAND 7
--------------------------------------------------------------
```

4.4 Working with Coordinates

When a shapefile is loaded, it is assigned to it a system of coordinates. A coordinate system is a reference system used to represent the locations of geographic features. This is either a geographic system, in which spherical coordinates are used to measure the distance of an object from the earth's centre, or a planimetric system, in which earth's coordinates are obtained by projecting objects onto a two-dimensional planar surface. The unit of measurement of coordinates depends on the chosen coordinate system. Typically we use latitude-longitude decimal degrees for geographic systems, and feet or metres for planimetric systems.

We can have a look to the coordinates of the countries contained in the shapefile by running Command (8), which returns the latitude and longitude coordinates for the first six countries contained in our map.

```
---------------------------------------------------------
RUN COMMAND 8
---------------------------------------------------------

> head(coord_countries)
[,1]          [,2]
5   17.572380 -12.3353410
18   2.343263   9.6476502
26  15.224274  -0.8400692
27  23.654965  -2.8760848
28  29.886825  -3.3562888
34  12.743335   5.6858605
```

Coordinates allow us to locate the position of an object on a map, and then to measure distances from other objects. This is what we do with Command (9), which returns a matrix of spatial distances between Angola, Benin and Congo. Specifically, each ijth cell indicates the distance in kilometres between the centroids of country i and country j.[4]

```
---------------------------------------------------------
RUN COMMAND 9
---------------------------------------------------------

> class(dist_cou)
[1] "matrix"
> dist_cou[1:3, 1:3]
       angola   benin    congo
angola 0.000    2970.773 1305.637
benin  2970.773 0.000    1844.2146
congo  1305.637 1844.215 0.0000
```

[4] A centroid is the average position of all the points of an object. Intuitively, it can be thought as the point at which a cutout of the object's shape could be perfectly balanced on the tip of a pin.

We can also create a "spatial point" object storing the coordinates of countries' centroid.

```
------------------------------------------------------------
RUN COMMAND 10
------------------------------------------------------------

> class(sp_coord_countries)
[1] "SpatialPoints"
attr(,"package")
[1] "sp"
```

R allows to assign a coordinate system also to the "spatial point" object. In this specific case, we use the World Geodetic System (WGS) 1984, which is a most popular geographic reference system.

```
------------------------------------------------------------
RUN COMMAND 11
------------------------------------------------------------
```

4.5 Modelling Spatial Interactions

Spatial interactions are usually modelled using an adjacency matrix, as those created with Commands (6) and (9). Depending on the form of interaction that we want to model, the ijth cell of the matrix can be a dummy variable indicating whether there is a connection between i and j or not (e.g. the matrix object created with Command 6). Alternatively, it can be a continuous variable in the realm of the positive numbers measuring the strength of interaction between the two countries (e.g. the matrix object created with Command 9).

4.5.1 Contiguity-Based Adjacency Matrices

The first form of spatial interaction discussed is that inferred from countries' spatial contiguity, i.e. i and j have a spatial connection if they are geographically contiguous. As it is shown in Fig. 4.2, three different forms of contiguity can be considered: "rook contiguity", which posits that two polygons are adjacent if they share an edge; "bishop contiguity", which assumes that polygons are adjacent if they have a common corner and "queen contiguity", which considers two polygons to be adjacent if they are rook or bishop contiguous.

Under this framework, we can construct a contiguity-based adjacency matrix where the ijth cell takes 1 if i and j are contiguous, i.e. they are neighbours, according to the chosen definition (i.e. rook, bishop or queen), and 0 otherwise. In order to do

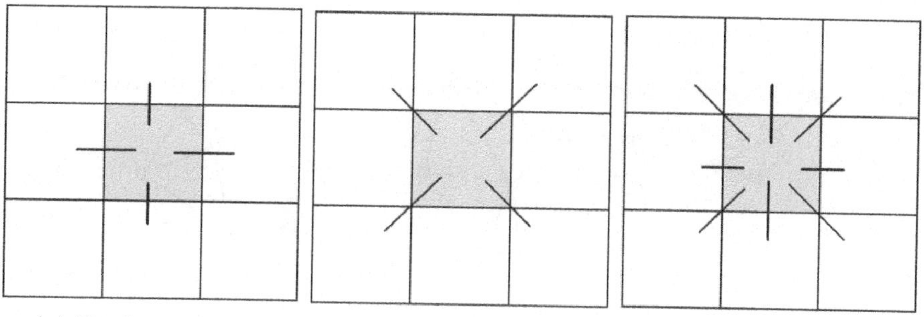

(a) Rook contiguity (b) Bishop contiguity (c) Queen contiguity

Fig. 4.2 Polygon contiguity

this in R, we use Command (12), which transforms the shapefile in a "nb" object, that is an object registering countries contiguity. Specifically, the command shows how to create an "nb" object using rook and queen contiguity.

```
---------------------------------------------------------------
RUN COMMAND 12
---------------------------------------------------------------
```

```
> summary(nb_coords_poly)
> summary(nb_coords_poly_rook)
```

The output of this command provides a number of relevant information. The "nb" object using queen contiguity ("nb_coords_poly") contains 43 countries. Each of them has on average 3 neighbours, and there are 150 pairs of contiguous countries. The output also shows the distribution of the number of neighbours in the data, e.g. five countries have no neighbours, two countries have one neighbour, and so on and so forth. The same logic applies when interpreting the information contained in the "nb" object using rook contiguity ("nb_coords_poly_rook"). It is worth to stress once again that more forms of interactions can be considered with queen contiguity than with rook contiguity. This motivates why the number of connections found in "nb_coords_poly" is higher than that obtained in the latter "nb_coords_poly_rook".

Once we have registered the list of each country's neighbours, we can create a contiguity-based adjacency matrix. This is done by transforming the "nb" object into a "listw" object using the function "nb2listw". When this is done, we can also apply some transformation to the matrix. For instance, suppose that each country has a limited amount of effort to dedicate to its partners, and that it chooses to interact with each one of them in the same way, i.e. the same amount of effort is assigned to each partner. This idea can be operationalized by row normalizing the matrix, so that each cell registers the percentage of effort designated to each partner, and the

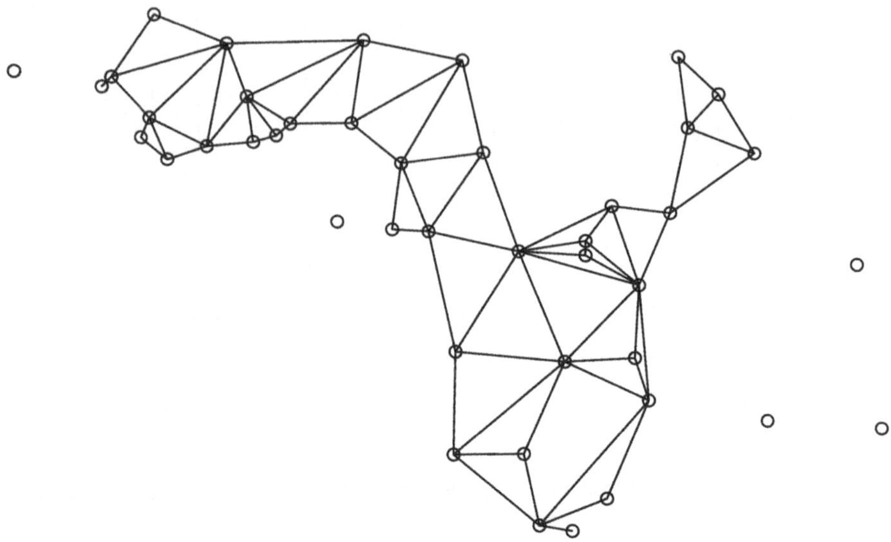

Fig. 4.3 Adjacent Sub-Saharan countries by Queen contiguity. *Note* each dot represents a country's capital city. Two dots are connected if they are adjacent to each other according to the Queen contiguity criterion

sum of all efforts is one. This is done by setting the field *style* = "*W*" in the function "nb2listw", as it is shown in Command (13).[5]

```
-------------------------------------------------------
RUN COMMAND 13
-------------------------------------------------------

> summary(nb_coords, zero.policy = TRUE)

> summary(unlist(nb_coords\$weights))
```

To visually inspect which countries are registered as adjacent in the "listw" object, one can run the Command (17), which returns Fig. 4.3. In the figure, each dot represents the centroid of a country, and two dots are connected if they are contiguous.

```
-------------------------------------------------------
RUN COMMAND 17
-------------------------------------------------------
```

[5] Alternatively, by setting: (i) *style* = "*B*", the weights of the adjacency matrix are transformed into the basic binary coding (Command 14), (ii) *style* = "*C*", the adjacency matrix is globally standardized, that is, it sums over all links to the number of considered areas, e.g. the number of countries in the shapefile (Command 15), (iii) *style* = "*U*", the adjacency matrix has weights equal to that of *style* = "*C*" divided by the number of considered areas (Command 16).

In order to have a look on the weights contained in the row-normalized adjacency matrix, we can transform the "listw" object to a "matrix" object using the function "listw2mat".[6] This is what Command (18) does, and in addition it returns the first three rows and columns of the matrix.

```
-----------------------------------------------------------
RUN COMMAND 18
-----------------------------------------------------------
```

```
> View(adj_mat[1:3, 1:3])
    AGO  BEN  BWA
AGO 0.00 0.00 0.25
BEN 0.00 0.00 0.00
BWA 0.20 0.00 0.00
```

4.5.2 Distance-Based Adjacency Matrices

A different form of spatial interaction that can be tracked is that obtained by looking at countries' distance. For example, one can hypothesize that spatial interactions occur under a specific threshold, e.g. i and j affect each other if they are located within 200 km from each other. To model such form of interaction, one can use the following commands. Specifically, replicating the approach of Sect. 4.5.1, we first create the "nb" object.

```
-----------------------------------------------------------
RUN COMMAND 20
-----------------------------------------------------------
```

```
> class(dnb200)
[1] "nb"
> dnb200
```

Then, we run the Command (21) to create a "listw" object registering spatial interactions within 200 km, as it is represented in Fig. 4.4.

```
-----------------------------------------------------------
RUN COMMAND 21
-----------------------------------------------------------
```

We now repeat the process by considering spatial interactions within 1000 km.

```
-----------------------------------------------------------
RUN COMMAND 22
-----------------------------------------------------------
```

[6] Note that to revert the process, i.e. transform a "matrix" object to a "listw" object, one can run Command (19).

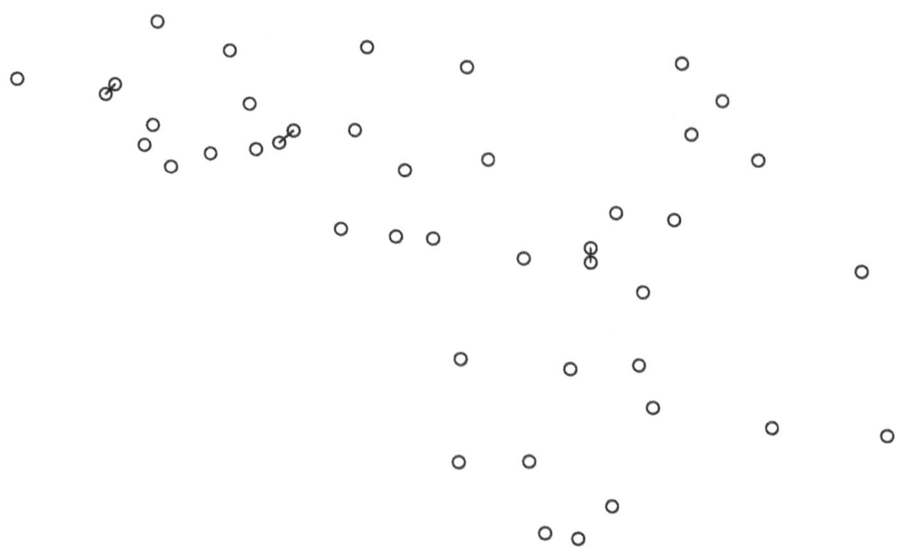

Fig. 4.4 Adjacent Sub-Saharan countries within 200 km. *Note* each dot represents a country's capital city. Two dots are connected if they lie within 200 km from each other

```
> class(dnb1000)
[1] "nb"
> dnb200

-----------------------------------------------------------
RUN COMMAND 23
-----------------------------------------------------------
```

The result of this exercise can be visualized in Fig. 4.5.

4.5.3 K-Nearest-Neighbour Weights Matrix

Another common hypothesis is that each spatial unit has a minimum number of interactions regardless of its distance from the other spatial units. This can be operationalized as follows. First, set a minimum number of interactions. We refer to this number as k, and we assume for the moment that $k = 1$, i.e. each country has at least one interaction. Second, for each country, find its nearest neighbour, and assume that there is a connection between the two. Third, measure the distance between each country and its neighbour, and take the maximum distance value registered. We refer to this value as d. Fourth, create an adjacency matrix where two countries are connected if they are located from each other within a distance radius lower than or

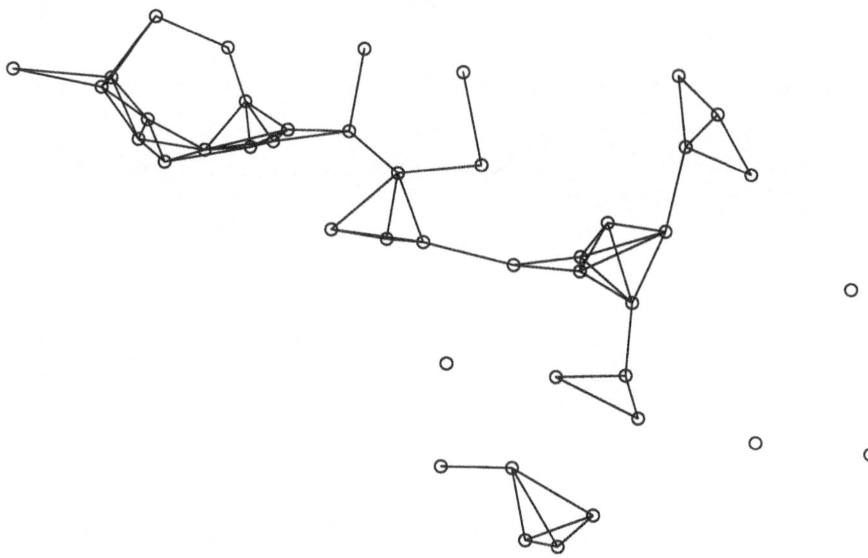

Fig. 4.5 Adjacent Sub-Saharan countries within 1000 km. *Note* each dot represents a country's capital city. Two dots are connected if they lie within 1000 km from each other

equal to d. The result of this process is a k-nearest-neighbour weights matrix, where each country has at least one neighbour.

The following commands show how to create a k-nearest-neighbour weights matrix with $k = 1$. The procedure mimics that presented in Sect. 4.5.2. First, we create a "nb" object using Command (24), where each country is connected to its nearest neighbour.

```
------------------------------------------------------------
RUN COMMAND 24
------------------------------------------------------------
```

Then, we run Command (25) which produces a "listw" object and returns the network of spatial interactions in Fig. 4.6.

```
------------------------------------------------------------
RUN COMMAND 25
------------------------------------------------------------
```

The distances between countries' centroid contained in the "listw" object can be extracted and examined using the function "nbdists", as shown in Command (26).

```
------------------------------------------------------------
RUN COMMAND 26
------------------------------------------------------------
```

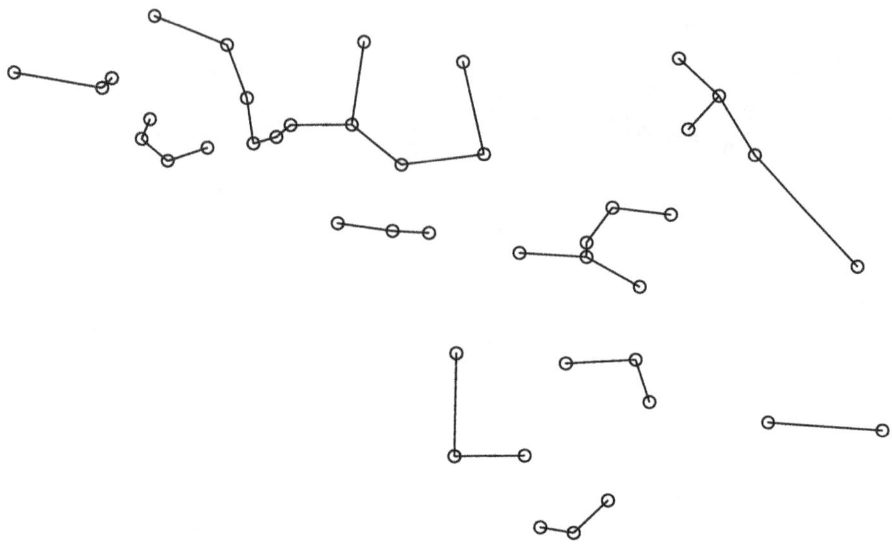

Fig. 4.6 K-nearest neighbours among Sub-Saharan countries: $k = 1$. *Note* each dot represents a country's capital city. Two dots are connected if they are adjacent according to the k-nearest-neighbour criterion, for $k = 1$

```
> k1.dist
[1] 1085.0252  194.2713  382.2910  694.7376  150.2350  669.8242  992.3344
[8]  862.2230  907.9619  474.0016  567.8490  474.0016  141.6248  382.2910
[15]  248.7733  226.4986  434.6789  614.5019  360.1733 1141.8833  593.1843
[22] 1141.8833  786.3573  466.9400  466.9400  877.6034  632.5126  150.2350
[29] 1595.8580  305.6141  305.6141  680.3945  141.6248  226.4986  725.1823
[36]  194.2713  585.0640  633.9022  454.8609  481.1923  680.3945  462.8260
[43]  705.6380
```

Now we run Command (27) to identify the maximum distance between two countries' centroid, i.e. 1595.858 km, that is, the threshold distance under which two countries are considered to have an interaction (i.e. d).

RUN COMMAND 27

```
> all.linkedT
[1] 1595.858
```

Finally, we run Command (28) to construct a k-nearest-neighbour weights matrix, where two countries are connected if their centroids are located within 1595.858 km.

RUN COMMAND 28

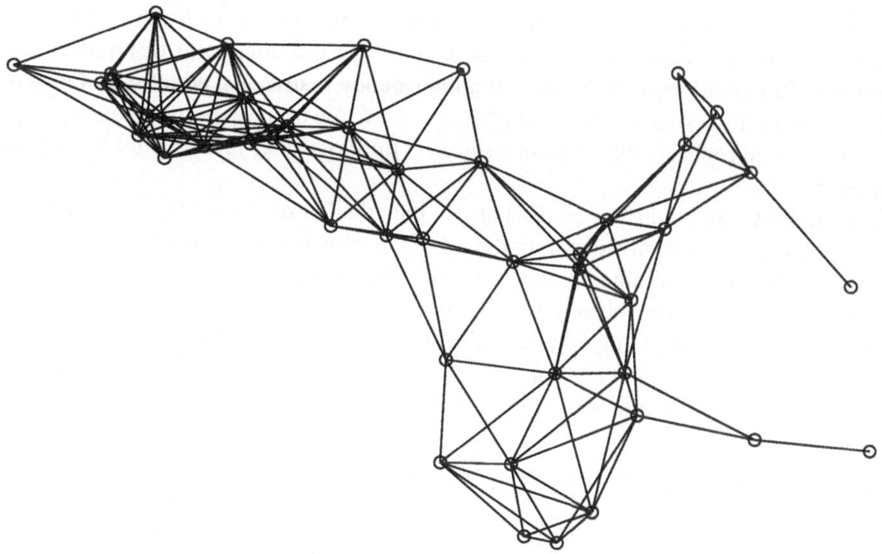

Fig. 4.7 All linked Sub-Saharan countries. *Note* each dot represents a country's capital city. Two dots are connected which lie within 1598.858 km from each other

The network of spatial interactions obtained in this way can be visually inspected by running Command (29), which returns Fig. 4.7.

```
----------------------------------------------------------
RUN COMMAND 29
----------------------------------------------------------
```

To determine the minimum and maximum number of countries' interactions obtained by this method one can run Command (30).

```
----------------------------------------------------------
RUN COMMAND 30
----------------------------------------------------------
```

```
> range(card(dnb.all.linkedT))
[1]  1 13
```

4.6 Modelling Social Interactions

4.6.1 Creating and Plotting an Edgelist

Social interactions are registered by the adjacency matrix created with Command (6), where country i is connected to country j if j represents a top 5 importers of country i. To match the information contained in this object with other data, it is necessary to transform it into an edgelist, that is, a different form of representing countries' interactions with respect to an adjacency matrix. An edgelist is a matrix composed of three vectors. The first indicates the name of the exporting country (e.g. i). The second registers the name of a top 5 i's importer (e.g. j). The third is a vector of ones.

Running Command (31), we first create the edgelist representing the network of top 5 import flows. Then we extract the set of coordinates associated to exporter and importer countries contained in the shapefile. Finally, we merge the coordinates with the edgelist. As a result, the edgelist now includes four new vectors. The first and the second contain, respectively, the latitude and longitude associated to the centroid of the exporter. The third and the fourth contain, respectively, the latitude and longitude associated to the centroid of the importer.

```
------------------------------------------------------------
RUN COMMAND 31
------------------------------------------------------------
```

We now plot the edgelist on the map. To do so, we need to use the R software package "ggplot2", which is able to combine the information contained in a shapefile with that stored in an edgelist. For this reason, we first transform the shapefile into an object that can be handled by "ggplot2".

```
------------------------------------------------------------
RUN COMMAND 32
------------------------------------------------------------
```

We then declare the visual settings of the plot that "ggplot2" has to draw. We also store the coordinates of a polygon containing the African continent. These will be used later to zoom in the map of the world and focus on the map of the Africa.

```
------------------------------------------------------------
RUN COMMAND 33
------------------------------------------------------------
```

We create the map of the world with Command (34), which returns Fig. 4.8.

```
------------------------------------------------------------
RUN COMMAND 34
------------------------------------------------------------
```

We zoom in the map returned by Command (34) using the set of coordinates declared in Command (33), and we obtain the map of the Sub-Saharan countries with Command (35), which returns Fig. 4.9.

```
------------------------------------------------------------
RUN COMMAND 35
------------------------------------------------------------
```

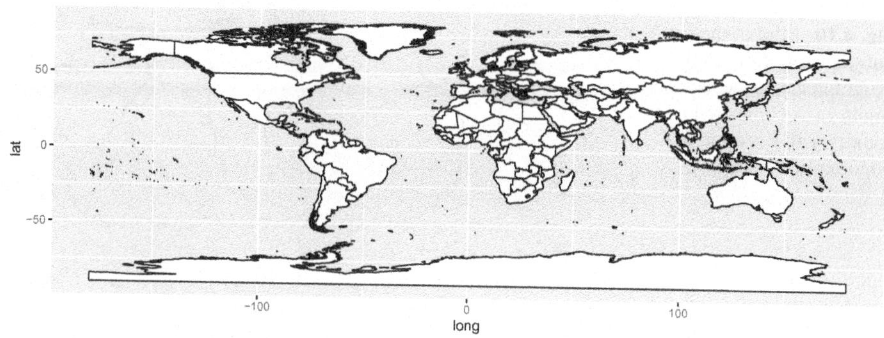

Fig. 4.8 Map of the World

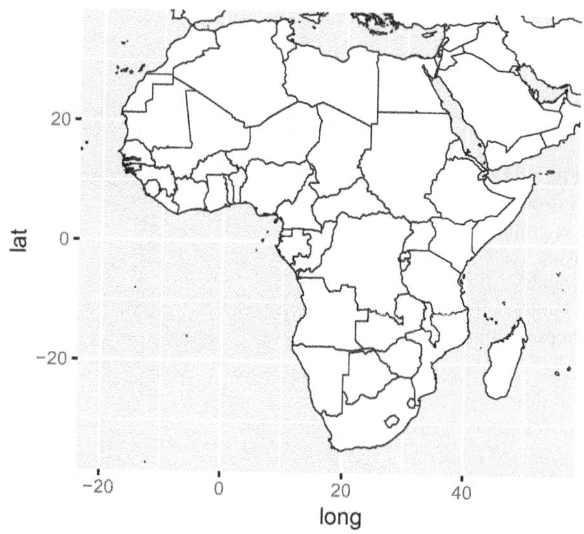

Fig. 4.9 Map of the Sub-Saharan countries

We can now plot network import flows on the map using Command (36), which returns Fig. 4.10.

```
-----------------------------------------------------------
RUN COMMAND 36
-----------------------------------------------------------
```

Finally, we add the visualization settings declared in Command (34) to produce Fig. 4.11.

```
-----------------------------------------------------------
RUN COMMAND 37
-----------------------------------------------------------
```

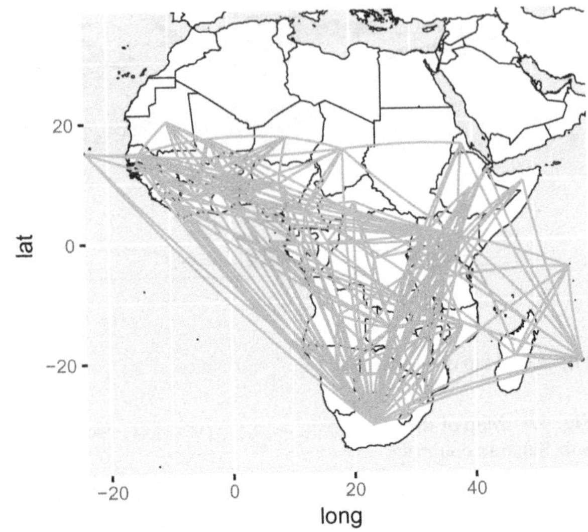

Fig. 4.10 Map of the top 5 Sub-Saharan flows in the geographical space. *Note* country i is connected to country j if j is a top 5 importer of country i

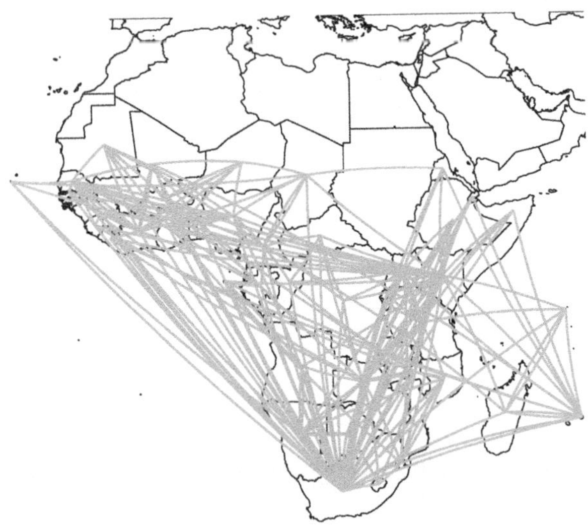

Fig. 4.11 Map of the top 5 Sub-Saharan flows in the geographical space (with fixed visual settings). *Note* country i is connected to country j if j is a top 5 importer of country i

4.6.2 *Creating and Plotting Network Data*

We can extrapolate a number of information from the top 5 import flow networks. This can be done by using the R software package "igraph", which allows us to use several network analysis techniques. To this end, we first transform the adjacency matrix created with Command (6) into an object which can be handled by "igraph".

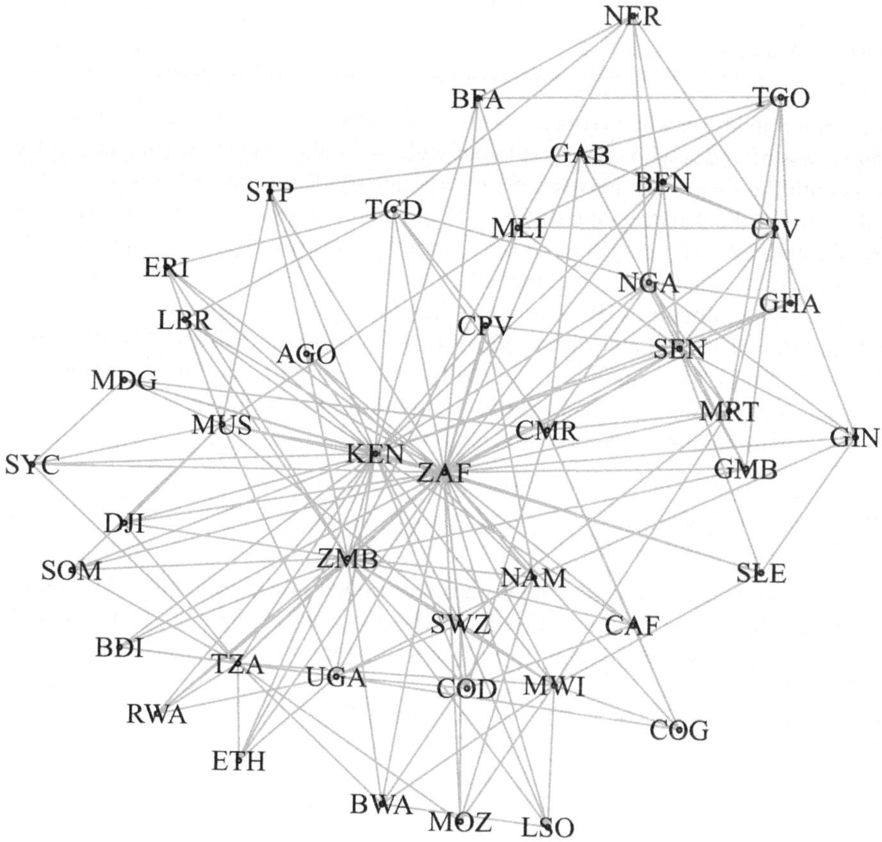

Fig. 4.12 Map of the top 5 Sub-Saharan flows in the social space. *Note* country i is connected to country j if j is a top 5 importer of country i. Country names are indicated using ISO3 code standard

```
--------------------------------------------------------
RUN COMMAND 38
--------------------------------------------------------
```

The network can be visualized using Command (39), which returns Fig. 4.12. The figure layout is produced by the Fruchterman and Reingold algorithm (Fruchterman and Reingold 1991), which is one of the most popular algorithms used to plot complex networks. Intuitively, the purpose of the algorithm is to visually stress peculiar patterns of interactions among nodes, by plotting the network so that nodes with shared connections are placed close to one another and far from other nodes.[7]

[7] Specifically, this is a force-directed layout algorithm which has the purpose to replicate a physical system in a mechanical equilibrium state. In this setting, nodes are represented by steel rings and the edges are springs between them. The attractive force is analogous to the spring force, and the

---

RUN COMMAND 39

A common analysis when working with social networks is to detect communities, that is, sets of nodes who densely interact with each other, but show sparser connections with the rest of the network (see Fortunato 2010 for a detailed review). This is done using the walktrap algorithm by Pascal and Latapy (2006), that is, a method to discover communities in a large-scale complex network. The idea is that random walks on a network tend to get "trapped" into densely connected set of nodes corresponding to communities. Recent researches show that the communities identified by random-walk-based algorithms are structurally close to real-world communities (Su et al. 2017). Communities obtained by this algorithm can be extrapolated by running Command (40). In our setting, communities can be interpreted as groups of countries with intense trade exchanges.

RUN COMMAND 40

We can visualize how network nodes are partitioned into communities by running Command (41), which returns Fig. 4.13. This is equal to Fig. 4.12, with the only difference that nodes belonging to the same community are displayed within the same coloured cloud.

RUN COMMAND 41

We can also visualize how communities are geographically distributed on the map. To this purpose, we need again to transform the data in a way that can be handled by the R software package "ggplot2". This is done with Command (42).

RUN COMMAND 42

We can now plot network communities, obtaining Fig. 4.14. Interestingly, we note that countries that are geographically close to one another tend also to belong to the same community in the network of top 5 import flows. This might be hinting to some spatial economic process, e.g. countries find easier to trade with other countries that are located in the nearest proximity. Such finding can also be relevant in policy terms. For instance, one can hypothesize that the occurrence of a drought reducing agro-food production in country i is most likely to deteriorate the consumption level of i's nearest neighbours, as these belong to i's network community and they are those most interconnected with i.

repulsive force is analogous to the electrical force. The aim of the algorithm is to minimize the energy of the system by moving the nodes and changing the forces between them.

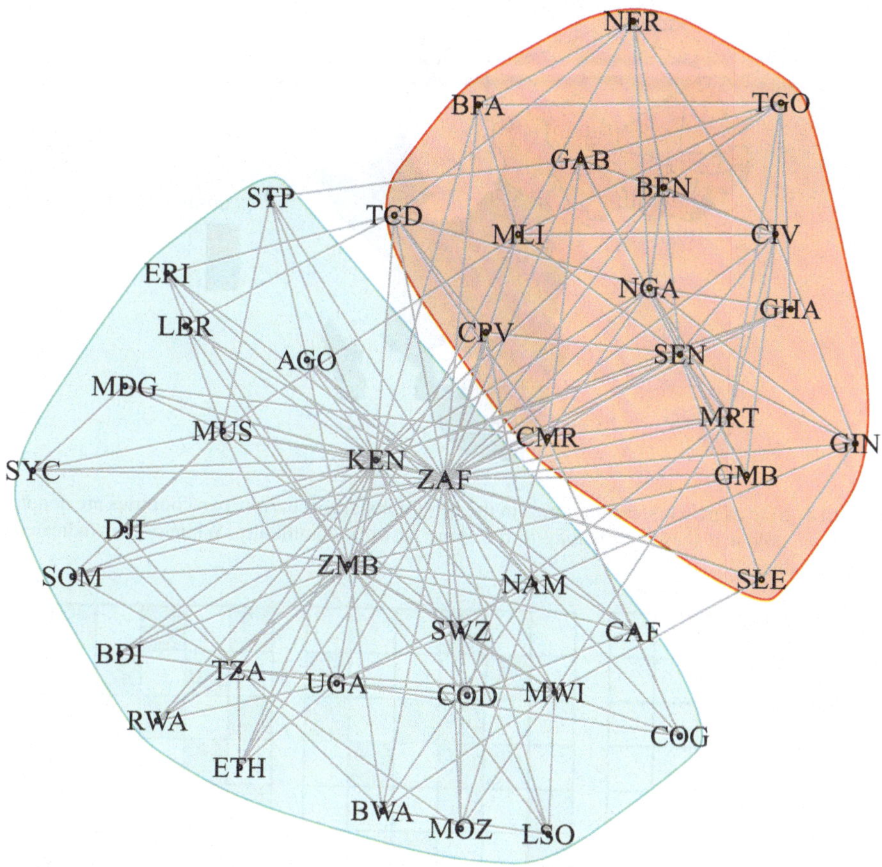

Fig. 4.13 Map of network communities in the social space. *Note* country i is connected to country j if j is a top 5 importer of country i. Two countries are within the same coloured region if they both belong to the same network community. Country names are indicated using ISO3 code standard

```
----------------------------------------------------------------
RUN COMMAND 43
----------------------------------------------------------------
```

4.7 Spatial Tests for Autocorrelation

Spatial tests for autocorrelation are used to assess the extent to which a spatial unit is similar to other nearby units and they indicate whether specific econometric estimation methods should be implemented to account for the spatial dimension when investigating unit's economic outcomes. Over the years, a number of tests have been

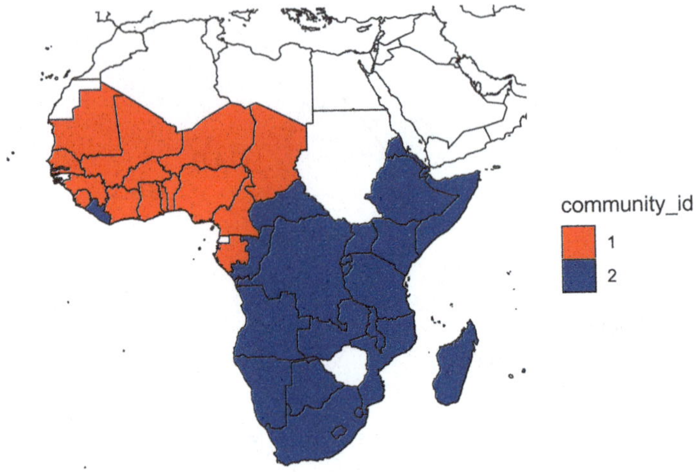

Fig. 4.14 Map of network communities in the geographical space. *Note* two countries are denoted with same colour if they both belong to the same network community. White colour indicates a missing value

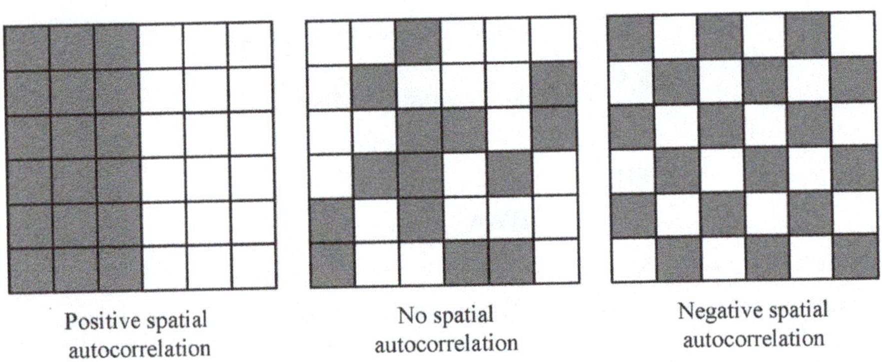

Fig. 4.15 Moran's I configuration

provided by the spatial econometric literature (see LeSage and Page 2009 for an extensive review). Here we will limit the discussion to four of the most popular tests, that is, the join count test (Cliff and Ord 1981), the global Moran's I test (Moran 1950), the local Moran's I test (Anselin and Getis 1992) and the Geary's c test (Geary 1954). Three different forms of spatial configurations can be analysed with such tests: positive spatial autocorrelation, where units with similar values are clustered in space; negative spatial autocorrelation, where units with similar values are dispersed in space and no presence of specific spatial patterns. Such configurations are exemplified in Fig. 4.15.

4.7.1 The Join Count Test

The join count test is used to assess the degree of clustering or dispersion of a categorical variable in the space. The reader interested in the details of this test can refer to Sect. (4.B.1). Here, in an effort to limit the discussion, we only provide the intuition of the test. Consider a binary variable mapped in two colours, e.g. Black (B) and White (W). Specifically, assume that the variable takes 1 when it refers to the colour black, and it takes 0 otherwise. Now let a join, or edge, be either WW (0-0), BB (1–1) or BW (1–0). A join count statistics indicates

- positive spatial autocorrelation (clustering), if the number of BB and WW joins is significantly higher than we would expect by chance.
- negative spatial autocorrelation (dispersion), if the number of BB and WW joins is significantly lower than we would expect by chance.
- null spatial autocorrelation, if the number of BB and WW joins is approximately the same as what we would expect by chance.

We use the join count test to analyse the extent to which the communities detected in the top 5 import flow networks (see Sect. 4.6.2) are geographically clustered. This is done with Command (44)

RUN COMMAND 44

```
> joincount.test(test, nb_coords, zero.policy = T)
```

The output of the test returned by R is interpreted as follows. The field "Expectation" indicates the expected number of joins, that is, the number of adjacent country pairs belonging to the same community. The field "Same colour statistic" reports the observed number of joins. Since the latter is higher than the former, we find evidence of positive spatial autocorrelation, i.e. countries belonging to the same community tend to cluster in space more than expected. Statistical significance of the test is indicated by the field "p-value".

4.7.2 The Global Moran's I Test

Moran's I statistics is a continuous variable going from -1 to 1. When approaching the former value, we have evidence of negative spatial autocorrelation (dispersion). By contrast, when approximating the latter value, data support the hypothesis of positive spatial autocorrelation. Values close to 0 indicates no presence of spatial autocorrelation. Further details are presented in Sect. (4.B.2).

We run a Moran's I test in order to check whether countries with similar levels of imports are located close to one another. If we assume that import is, for example, a proxy of a country dependence from the agro-food goods produced abroad, finding evidence of positive spatial autocorrelation would be useful to identify highly interconnected areas in which national consumption might be particularly exposed to shocks in the trade network. To implement our test, we need the vector of import values created with Command (7), and to transform the shapefile of the map of Sub-Saharan countries into a "listw" object using Command (45).

RUN COMMAND 45

The test is then implemented running Command (46).

RUN COMMAND 46

```
> moran.test(sub_shp@data$import, listw = nb_x, randomisation=TRUE, zero.policy=TRUE)
```

The results show that there is a (mild) positive spatial autocorrelation in the location of countries with similar import values. In fact, the value of the I statistic (see the field "Moran I statistic") is positive and statistically significant (see the field "p-value").[8]

4.7.3 The Global Geary's c Test

Geary's c statistic is a measure of spatial autocorrelation. It is inversely correlated to Moran's I statistic (Anselin 1995); however, the two are not identical. The former is more sensitive to local patterns of autocorrelation, while the latter is better suited for detecting autocorrelation at the global level.

The interpretation of the statistic is the following: (i) when there is no autocorrelation present, $c \cong 1$; (ii) when there is a maximum positive autocorrelation present $c \cong 0$, when there is a maximum negative autocorrelation present $c \cong 2$.[9] The test is run using Command (47).

RUN COMMAND 47

```
> geary.test(sub_shp@data\$import, listw = nb_x, randomisation=TRUE, zero.policy=TRUE)
```

The interpretation of the output mimics that used for discussing Command (46). The value of c is lower than 1, and hence it confirms the presence of a (mild) positive spatial autocorrelation found in Sect. 4.7.2.

---

[8]For further details on the output of this command, we suggest the reader to run the script "?moran.test" in the R console.

[9]For further details, see Sect. (4.B.3).

4.7.4 Local Moran's I Test

Anselin (1995) has shown that Moran's I spatial autocorrelation statistic (Sect. 4.7.2) can be decomposed into local values.[10] This allows to test the extent to which each unit, say a country, is spatially autocorrelated with the surrounding units.

We implement the test to complement the analysis in Sect. 4.7.2 by running Command (48), which returns, for each considered country, the value of the local Moran's I, its variance (field "Var.Ii") and the value of Moran'sI test (field "Z.Ii") and its statistical significance (field "Pr(z>0)").

```
-----------------------------------------------------------
RUN COMMAND 48
-----------------------------------------------------------
```

```
> head(LM_Results)
```

The information produced by Command (48) can be summarized using a Moran scatterplot. To this purpose, we first prepare data to be plotted using the above-mentioned R software package "ggplot2".

```
-----------------------------------------------------------
RUN COMMAND 49
-----------------------------------------------------------
```

The data obtained from Command (49) are then used to produce the Moran scatterplot in Fig. 4.16. The interpretation of a Moran scatterplot is the following. The x-axis indicates the import values of country i and the y-axis indicates the average import value of the countries adjacent to i. On the left quadrants are positioned countries with low import values which are close to countries with either low import values (bottom-left quadrant) or high import values (upper-left quadrant). On the right quadrant are positioned countries with high import values which are close to countries with either low import values (bottom-right quadrant) or high import values (upper-right quadrant). The position of the countries in the quadrant is used to identify "clusters" and "hotspots". A cluster is a group of contiguous countries located alternatively on the upper-right quadrant or on the bottom-left quadrant. A hotspot is a group of contiguous countries located alternatively on the upper-left quadrant or on the bottom-right quadrant.

```
-----------------------------------------------------------
RUN COMMAND 50
-----------------------------------------------------------
```

The information contained in the Moran scatterplot can also be used to create the map in Fig. 4.17. This is done by assigning the colour blue to the countries located on the left quadrants of the plot, and the colour red to the countries positioned on the right quadrants of the plot. Dark hues of colours are used to identify clusters, which

[10]For further details, see Sect. (4.B.4).

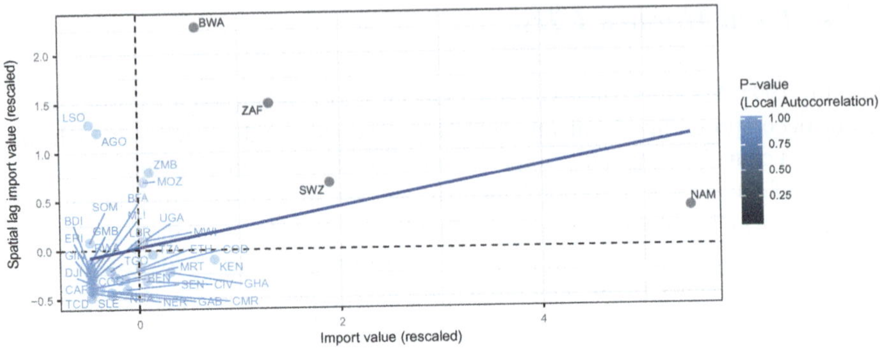

Fig. 4.16 Moran's I scatterplot. *Note* X-axis registers the import value of each country i. Y-axis indicates the average import value of i's neighbours. Colour intensity is used to indicate the statistical significance detected when looking at the spatial autocorrelation between a country import value and that of its neighbours, i.e. the darker the colour, the stronger evidence of statistically significant spatial autocorrelation. A regression line is included. Country names are indicated using ISO3 code standard

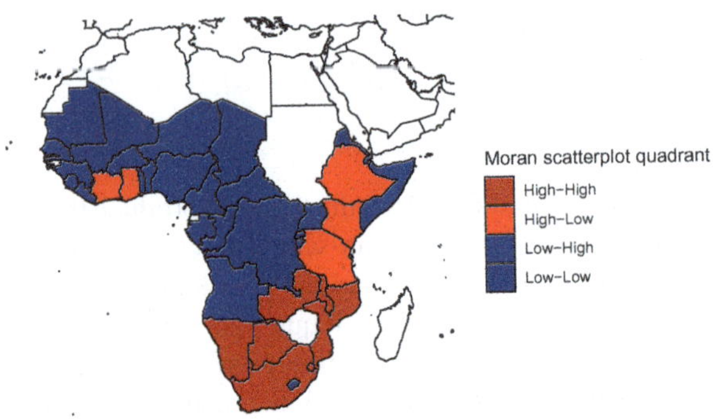

Fig. 4.17 Moran's I map. *Note* countries in dark red are those located in the upper-right quadrant of Fig. 4.16. Countries in light red are those located in the bottom-right quadrant of Fig. 4.16. Countries in dark blue are those located in the bottom-left quadrant of Fig. 4.16. Countries in light blue are those located in the upper-left quadrant of Fig. 4.16. White colour indicates a missing value

are associated either to dark red or deep blue. By contrast, lighter colours indicate hotspots.

In order to produce the map, we first need to reshape the data in a format that can be handled by "ggplot2".

```
RUN COMMAND 51
```

Then we can plot the map using Command (52).

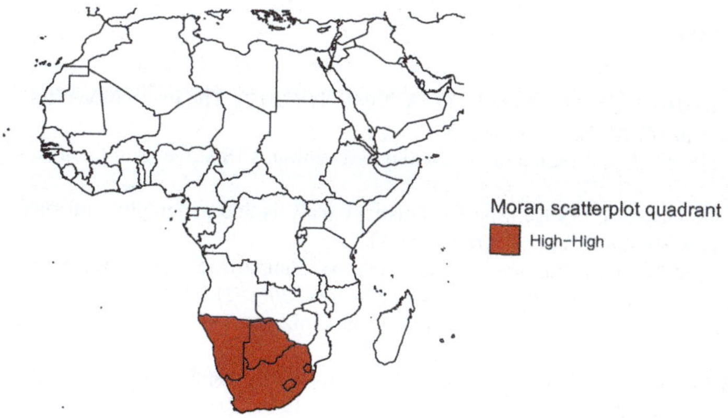

Fig. 4.18 Moran's I map—significant autocorrelation. *Note* countries in dark red are those located in the upper-right quadrant of Fig. 4.16. Countries in light red are those located in the bottom-right quadrant of Fig. 4.16. Countries in dark blue are those located in the bottom-left quadrant of Fig. 4.16. Countries in light blue are those located in the upper-left quadrant of Fig. 4.16. Colour is reported only for those countries featuring spatial autocorrelation in their import value at the 5% significant level

```
-------------------------------------------------------------
RUN COMMAND 52
-------------------------------------------------------------
```

We can also isolate the countries for which the local Moran's *I* test returned a statistically significant level of autocorrelation at the 5% level, assigning a colour only to them. This is done with Command (53), which returns Fig. 4.18.

```
-------------------------------------------------------------
RUN COMMAND 53
-------------------------------------------------------------
```

Interestingly, the figure shows that there is a cluster with a statistically significant level of autocorrelation in the area surrounding South Africa.

4.8 Conclusions

In this chapter, the reader is introduced to a number of routines that can be run in R to model spatial, economic and social interactions among spatial units using GIS methods. In addition, the reader is introduced to the most popular statistics to test the level of spatial autocorrelations in the values featured by units. This chapter represents, therefore, a useful toolbox for the researcher interested in learning the basics for manipulating GIS data with R, and in approaching the study of economics from a spatial perspective by using GIS-produced data.

References

Abbate, A., De Benedictis, L., Fagiolo, G., & Tajoli, L. (2012). The international trade network in space and time. *LEM Papers Series* 17.

Anselin, L. (1995). Local indicators of spatial association, LISA. *Geographical Analysis*, *27*(2), 93–115.

Anselin, L., & Getis, A. (1992). Spatial statistical analysis and geographic information systems. *The Annals of Regional Science*, *26*(1), 19–33.

Aten, B. H. (1997). Does space matter? International comparisons of the prices of tradables and nontradables. *International Regional Science Review*, *20*(1–2), 35–52.

Baum-Snow, N. (2007). Did highways cause suburbanization? *Quarterly Journal of Economics*, *122*, 775–805.

Baybeck, B., & Huckfeldt, R. (2002). Urban contexts, spatially dispersed networks, and the diffusion of political information. *Political Geography*, *21*(2), 195–220.

Bevc, C. A., Barlau, A. N., & Passanante, N. A. (2009). Mapping convergence points in the initial emergency response to 9/11 1. *Disasters*, *33*(4), 786–808.

Bivand, R., & Lewin-Koh, N. (2019). *MAPTOOLS: Tools for handling spatial objects*, r package version 0.9-5.

Bivand, R., & Rundel, C. (2018). *RGEOS: Interface to Geometry Engine - Open Source ('GEOS')*, r package version 0.4-2.

Bivand, R. S., Pebesma, E., & Gomez-Rubio, V. (2013). *Applied spatial data analysis with R* (2nd ed.). NY: Springer.

Bivand, R., Keitt, T., & Rowlingson, B. (2019). *RGDAL: Bindings for the 'geospatial' data abstraction library*, r package version 1.4-3.

Burgess, R., Hansen, M., Olken, B. A., Potapov, P., & Sieber, S. (2012). The political economy of deforestation in the tropics. *The Quarterly Journal of Economics*, *127*(4), 1707–1754.

Butts, C. T., Acton, R. M., Hipp, J. R., & Nagle, N. N. (2012). Geographical variability and network structure. *Social Networks*, *34*(1), 82–100.

Chandra, A., Hajra, K., Kumar, D. P., & Sen, P. (2007). Modeling temporal and spatial features of collaboration network. *International Journal of Modern Physics, C*, *18*, 1157–1172.

Chen, Y., Jin, G. Z., Kumar, N., & Shi, G. (2013). The promise of Beijing: Evaluating the impact of the 2008 Olympic Games on air quality. *Journal of Environmental Economics and Management*, *66*(3), 424–443.

Cheshire, P., & Sheppard, S. (1995). On the price of land and the value of amenities. *Economica*, *62*, 247–267.

Church, R., & Murray, A. (2009). *Business site selection, location analysis and GIS*. New York: Wiley.

Clarke, K. (2003). *Getting started with geographic information systems*. Upper Saddle River, NJ: Prentice Hall.

Cliff, A. D., & Ord, J. K. (1981). *Spatial processes*. London, UK: Pion.

Csardi, G., & Nepusz, T. (2006). The igraph software package for complex network research. *InterJournal Complex Systems*, *1695*, 1–9.

Dell, M. (2010). The persistent effects of Peru's mining mita. *Econometrica*, *78*(6), 1863–1903.

Dell, M. (2015). Trafficking networks and the Mexican drug war. *American Economic Review*, *105*(6), 1738–1779.

Dinkelman, T. (2011). The effects of rural electrification on employment: New evidence from South Africa. *American Economic Review*, *101*(7), 3078–3108.

Donaldson, D., & Hornbeck, R. (2016). Railroads and American economic growth: A "market access" approach. *The Quarterly Journal of Economics*, *131*(2), 799–858.

Duflo, E., & Pande, R. (2007). Dams. *The Quarterly Journal of Economics*, *122*, 2,601–2,646.

Durante, R., Pinotti, P., & Tesei, A. (2019). Political legacies of Italian entertainment TV. *American Economic Review*, *109*(7), 2497–2530.

Festinger L., Schachter S., & Back, K. (1950). *Social pressures in informal groups; a study of human factors in housing*. New York City, NY: Harper.

Fortunato, S. (2010). Community detection in graphs. *Physics Reports*, *486*(3–5), 75–174.

Fruchterman, T., & Reingold, E. (1991). Software: Practice and experience. *Quarterly Journal of Economics*, *21*(11), 1129–1164.

Geary, R. C. (1954). The contiguity ratio and statistical mapping. *The Incorporated Statistician*, *5*(3), 115–146.

Henderson, J. V., Storeygard, A., & Weil, D. N. (2012). Measuring economic growth from outer space. *American Economic Review*, *102*(2), 994–1028.

Henry L., & Wickham, H. (2019). *PURRR: Functional Programming Tools*, r package version 0.3.2.

Hijmans, R. J. (2017). *Geosphere: Spherical trigonometry*, r package version 1.5-7.

Hijmans, R. J. (2019). *Raster: Geographic data analysis and modeling*, r package version 2.8-19.

Hodler, R., & Raschky, P. A. (2014). Regional favoritism. *The Quarterly Journal of Economics*, *129*(2), 995–1033.

Jackson, M. (2010). *Social and Economic Networks*. Princeton (NJ), USA: Princeton University Press.

Jackson, M., & Zenou, Y. (2013). *Economic analyses of social networks volume I: Theory*. Cheltenham, UK: Edward Elgar Publishing.

Jackson, M. O., Rogers, B. W., & Zenou, Y. (2017). The economic consequences of social-network structure. *Journal of Economic Literature*, *55*(1), 49–95.

Kahle, D., & Wickham, H. (2013). GGMAP: Spatial visualization with ggplot2. *The R Journal*, *5*(1), 144–161.

Koskinen, J., & Lomi, A. (2013). The local structure of globalization. *Journal of Statistical Physics*, *151*(3–4), 523–548.

LeSage, J., & Pace, R. K. (2009). *Introduction to spatial econometrics*. Chapman and Hall/CRC.

Lomi, A., & Pallotti, F. (2012). Relational collaboration among spatial multipoint competitors. *Social Networks*, *34*(1), 101–111.

Longley, P., Goodchild, M., Maguire, D., & David, R. (2005) *Geographical information systems and science*. Wiley.

McPherson, M., Smith-Lovin, L., & Cook, J. M. (2001). Birds of a feather: Homophily in social networks. *Annual Review of Sociology*, *27*(1), 415–444.

Michaels, G. (2008). The effect of trade on the demand for skill - evidence from the interstate highway system. *Review of Economics and Statistics*, *90*, 683–701.

Michalopoulos, S., & Papaioannou, E. (2013). National institutions and subnational development in Africa. *The Quarterly Journal of Economics*, *129*(1), 151–213.

Michalopoulos, S., & Papaioannou, E. (2016). The long-run effects of the scramble for Africa. *American Economic Review*, *106*(7), 1802–1848.

Moran, P. A. P. (1950). Notes on continuous stochastic phenomena. *Biometrika*, *37*, 1/2, 17–23.

Mouw, T., & Entwisle, B. (2006). Residential segregation and interracial friendship in schools. *American Journal of Sociology*, *112*(2), 394–441.

Nunn, N. (2008). The long-term effects of Africa's slave trades. *The Quarterly Journal of Economics*, *123*(1), 139–176.

Olken, B. A. (2009). Do television and radio destroy social capital? Evidence from Indonesian villages. *American Economic Journal: Applied Economics*, *1*(4), 1–33.

Overman, H. G. (2008). GIS data in economics, The New Palgrave Dictionary of. *Economics*, *1–8*, 2478–2480.

Pascal, P., & Latapy, M. (2006). Computing communities in large networks using random walks. *Journal of Graph Algorithms and Applications*, *10*(2), 191–218.

Pebesma, E. (2018). Simple features for R: Standardized support for spatial vector data. *The R Journal*.

Pebesma, E. J., & Bivand, R. S. (2005). Classes and methods for spatial data in R. *R News*, *5*(2), 9–13.

Pedersen, T. L. (2018). *GGRAPH: An implementation of grammar of graphics for graphs and networks*, r package version 1.0.2.
Preciado, P., Snijders, T. A. B., Burk, W. J., Stattin, H., & Kerr, M. (2012). Does proximity matter? Distance dependence of adolescent friendships. *Social Networks*, *34*(1), 18–31.
Qian, N. (2008). Missing women and the price of tea in China: The effect of sex-specific earnings on sex imbalance. *The Quarterly Journal of Economics*, *123*(3), 1251–1285.
Rogall, T. (2015). Mobilizing the masses for genocide.
Slowikowski, K. (2018). *GGREPEL: Automatically position non-overlapping text labels with 'ggplot2'*, r package version 0.8.0.
Storeygard, A. (2016). Farther on down the road: Transport costs, trade and urban growth in sub-Saharan Africa. *The Review of Economic Studies*, *83*(3), 1263–1295.
Su, Y., Wang, B., & Xingyi, Z. (2017). A seed-expanding method based on random walks for community detection in networks with ambiguous community structures. *Scientific Reports*, *7*, 41830.
Takhteyev, Y., Gruzd, A., & Wellman, B. (2012). Geography of Twitter networks. *Social Networks*, *34*(1), 73–81.
Warnes, A. M. (1986). The residential mobility histories of parents and children, and relationships to present proximity and social integration. *Environment and Planning A*, *18*(12), 1581–1594.
Wasserman, S., & Faust, K. (1994). *Social network analysis: Methods and applications*. Cambridge, UK: Cambridge University Press.
Wickham, H. (2011). The split-apply-combine strategy for data analysis. *Journal of Statistical Software*, *40*(1), 1–29.
Wickham, H. (2016). *ggplot2: Elegant graphics for data analysis*. New York: Springer.
Wickham, H., & Henry, L. (2019). *TIDYR: Easily tidy data with 'spread()' and 'gather()' functions*, r package version 0.8.3.
Wickham, H., Francois, R., Henry, L., & Muller, K. (2019). *DPLYR: A grammar of data manipulation*, r package version 0.8.0.1.
Yanagizawa-Drott, D. (2014). Propaganda and conflict: Evidence from the Rwandan genocide. *The Quarterly Journal of Economics*, *129*(4), 1947–1994.
Zipf, G. K. (1949). *Human behavior and the principle of least effort*. Cambridge, MA: Addison-Wesley Press.

Chapter 5
Resilience to Climate Change: Spatial Ricardian Analysis

Abstract This chapter provides practical and theoretically founded indications concerning the spatial econometric assessment of climate impacts, focusing specifically on the case of agriculture. The chapter, after introductory considerations, starts by dealing with measurement issues about climate change and weather extremes, and related suggestions about sample selection and the choice of the weather indicators. The Ricardian approach to climate resilience of the agricultural sector is then introduced, along with motivations behind a spatial Ricardian approach that allows for spatial autocorrelation. Modelling strategy issues are touched upon, concerning model selection and the spatially weighted matrix. Next, a review of the spatial Ricardian approach describes the key works that estimated spatial spillovers of weather events. A critical discussion closes the chapter. The discussion focuses on the information content that econometric estimates can provide on adaptation practices, as well as on the possible integration between spatial econometrics and agent-based computational projections of the climate-economy nexus.

Keywords Climate change · Weather extremes · Land value · Ricardian approach · Spatial econometrics · Adaptation · Damage function

5.1 Introduction

Much attention around resilience in economics has been sparked by the Great Recession and by decisions to leave the European Union (Brexit in Fingleton 2018 and even the possible "Italexit" in Fingleton 2019). Disruptive as it was, the Great Recession wiped out the belief on the "virtual disappearance of the business cycle" (Romer 1999) and served as a cautionary tale against "linear" views of the macroeconomy that dominate policy-making. Yet, in the perspective of making regional outputs and employment resilient to macroeconomic shocks, policy-makers can rely on wide

R. De Siano et al., *Regional Resilience to Climate and Environmental Shocks*, SpringerBriefs in Regional Science, https://doi.org/10.1007/978-3-030-54588-8_5

econometric evidence and on a large repertoire of fiscal, monetary, labour and trade policy tools employed by governments in the past.[1]

Not quite the same can be said about climate-related shocks. Humans in the past have learned to cope with extreme precipitation events, heat waves, windstorms and other weather extremes. Yet, in no period of human history after the industrial revolution can one record a comparable pace of climatic change as the one we are experiencing today. Resilience to climate change is resilience to outcomes of a non-linear dynamic relationship between climate and the economy, whose feedbacks from the economy to climate have only become visible in the most recent decades. In the pre-industrial era, the challenge for farmers was to cope with extreme weather events drawn from a probability distribution that could be considered stable over the course of a lifetime. Historical records were helpful in estimating the likelihood and timing of adverse weather conditions.[2] The challenge today is to cope with events from a probability distribution that will change in the future if economies do not depart from their "business as usual".

Radical uncertainty surrounding the future climate-economy nexus calls for a policy perspective that emphasizes flexibility in action and the ability of public programmes to fit the specificities of local communities and contexts. Policy and governance actors at various scales, including the regional one, should coordinate towards mitigation of the global warming externality, but in that process, interactions among local actors need to be carefully taken into account. Relatedly, (Burke et al. 2016) stressed that the rigorous quantification of indirect effects from climate-induced damages is of paramount importance, moving beyond first-order estimates.

Spatial analysis is appropriate in the understanding of climate impacts for a number of reasons:

- Regions are heterogeneous as regards their exposure to climate risks, in terms of infrastructures, asset portfolios, knowledge bases, land use, ecosystems, cultural heritage and population.
- Regions differ also in terms of political value systems and organization capabilities to evaluate risks, prepare for an adverse event and deal with its aftermath.
- Climate policies formulated upscale (e.g. at the EU level) may underperform if policy interests upscale and downscale (e.g. in regions) are not aligned, or if upscale actors do not have enough knowledge on regional specificities that may hamper policy implementation. Relatedly, the scale of resilience analysis needs to match the scale of decision-making of policy actors and other stakeholders (Schröter et al. 2005).
- Climate shocks propagate through space, first of all because of physical spillovers. Propagation is facilitated by geography, urbanization, interconnectivity (transport infrastructures), trade relationships, income differentials (that may stimulate

[1] As a matter of fact, the way shocks such as Brexit and Italexit have been modelled by Fingleton are formally indistinguishable from any other restrictive trade policy.

[2] Howard and Sterner (2017) report (in their footnote 2) some revealing statements on climate uncertainty by such authoritative scholars as Pindyck and Weitzman.

migration flows) and shared natural resources (e.g. a region's river basin being fed by another region's melting glacier).

- The speed of spatial propagation of climate shocks depends on the spread of information and the transmission of knowledge about adaptation practices. If space was immaterial with respect to knowledge transmission, technology should quickly spread to all locations. If so, climate impacts could be equivalently assessed by means of longitudinal analysis (tracking one location over time) or cross-sectional analysis (multiple locations for a single point in time). In other words, in that case the system is ergodic, and space and time are substitutable (Polsky 2004). Yet, knowledge tacitness (Polanyi 1966) implies that spatial proximity matters for the transmission of knowledge. The features of a neighbouring location are not irrelevant with regards to a location's resilience to shocks. Results concerning a given location are externally valid only for other locations that are similar in all respects, including the average features of their neighbours.

Given the above considerations, the problem of resilience to climate change poses specific challenges that call for a methodological discussion on the role of space in climate-economy interactions. In particular, we will focus on the spatial econometric approach to the estimation of what climate economists call the "damage function". A wide array of economic damages from extreme weather events can be envisioned, and nearly as many have been examined in the climate econometric literature. Carleton and Hsiang (2016) offer a comprehensive detour on estimated climate impacts on a wide array of economic and social outcomes, from income to health, from agriculture to conflict, while further highly cited reviews of climate econometrics are in Dell et al. (2014), Hsiang (2016). We have chosen to restrict our discussion and literature review to a rather specific case study, namely, agriculture.

Agriculture is among the most sensitive sectors to weather conditions and, despite it accounts for a limited share of value added and employment in advanced economies, it is the source of critical food resources for less developed countries where hunger and famine are still cogent problems. Furthermore, farming is not merely sensitive to weather conditions, it is affected by the spatial propagation of weather anomalies. The above example of the glacier-river basin relationship is a case in point. And the effects of climate change on land values are challenging in the perspective of distributive equity, as through adaptation, farmers in some geographical areas may benefit from the changing climate. The econometric literature has not yet reached consensus on the sign of the country wide effects of climate change on agriculture, at least for high-income countries such as the US, whereas there is no disagreement on the prediction that not all regions will be losers. As long as regional outputs are spatially correlated, spatial clusters may emerge in terms of income and wealth, causing concerns that policy-makers are expected to tackle.

The next sections provide practical but theoretically founded indications concerning the spatial econometric assessment of climate impacts. Section 5.2 deals with issues about measurement of weather and climate, given their definitions, and given the temporal and spatial limitations in data collection. On these premises, Sect. 5.3 introduces the spatial econometric approach to climate resilience as an extension of the Ricardian approach, according to which the correlation between economic outcomes and weather events emerges because agents include weather conditions as parameters in their decisions. The most significant applications of such a spatial Ricardian approach are reviewed in Sect. 5.4, where the focus is on works that estimated spatial spillovers of weather events. A critical discussion is provided in Sect. 5.5 based on the nexus between resilience, adaptation and mitigation, as well as on advances in integrated assessment modelling.

5.2 Measurement Issues

The design of an econometric analysis of the climate-economy nexus starts with some key decisions on how to translate generic categories (economic outcomes, climate anomalies) into operational measurements.

The first and foremost measurement challenge concerns climate variables (the explanatory variables in the analysis). Climate can be defined as the probability distribution of a set of meteorological variables, such as temperature, humidity, atmospheric pressure, wind and precipitation. Weather at a given time and location is understood as a draw from the probability distribution identified as climate. Similar definitions can be found in Dell et al. (2014), Hsiang (2016). By definition, weather is variable over time, even when observed at a very high frequency. Conversely, any change in climate corresponds to a slow shift in the moments of the weather distribution.[3] At any time, both climate and (therefore) weather are variable across space.

The existing climate econometric literature provides guidance on the choice of the weather events and measurements that are best included in a regression model, given the goal of the analysis.

First, it is useful to focus on weather events that have sparked significant concerns in recent times (e.g. the Australian wildfires). Second, the current exposure or coping ability of a region could be relevant in sample selection, but should be ideally balanced with considerations about which regions are expected to face an increasing exposure. Projections from international organizations such as the IPCC, for instance, based on integrated assessment models, may be useful in this (see Sect. 5.5).

Third, climate-related explanatory variables in regressions should take account of the skewed and heavy-tailed distributions of weather events and recognize the importance of climate variability on triggering natural disasters (Katz and Brown 1992; Renton et al. 2014; Thomalla et al. 2006; Revesz et al. 2014). Averages may be less informative than specific quantiles of a weather variable distribution.

---

[3] It is well known that climate has always changed in the course of history. Behringer (2010) contains a popular survey of methods through which climate historians and scientists reconstruct the dynamics of climate in history. A guide to climate science for economists has been compiled by Hsiang and Kopp (2018).

Fourth, there may be temporal and spatial correlation among different weather events, since many weather extremes are jointly caused by the warming climate. Multicollinearity is, therefore, to be avoided. At the same time, including only temperature-related variables does not allow to capture effects that are directly imputable to other weather events (e.g. changing wind patterns alter regional potentials for wind power generation; less snow hampers the inflow of tourists on the Alps).

Finally, climate econometric estimates are exposed to the full range of potential estimation biases, partly due to measurement issues: (i) measurement errors, due to imperfect knowledge on the statistical properties of weather; (ii) omitted variables, e.g. because the most vulnerable communities are often those with poorer data availability and quality; (iii) reverse causation, as we have imperfect knowledge on the time scale over which climate-economy feedbacks begin to unfold and (iv) selection biases, as human settlements and the sectoral composition of outputs reflect the climate conditions that had (relatively) stabilized over the last centuries. Related to the omitted variables problem, spatial effects can make estimates biased and inefficient (see Sect. 5.3).

Another major issue is that climate anomalies are not randomly assigned across agents, regions and countries. In this respect, Hsiang et al. (2019) provide an overview of the existing results on the exposure to environmental damages across space and income levels, confirming some expectations while upsetting some preconceptions. Extreme weather events considered in the review are extremely hot temperatures (e.g. greater than 30 °C), tornadoes, tropical cyclones and rainfalls. Some interesting take-home messages can be drawn from their analysis. One is that, albeit high-income economies are better equipped to face the adverse consequences of extreme weather, exposure as such is less dependent on per capita income than on geography. For instance, low-income countries are mostly localized in world regions where tornadoes are less likely to occur (strong temperature and pressure gradients). Another insightful evidence points to within-country correlations between per capita income and weather, i.e. poor populations tend to live in hotter and drier locations, although with some exceptions.

A common difficulty of climate econometrics, though, concerns the interpretation of the estimated marginal effects. One side of this difficulty is that retrospective estimates of the climate-economy nexus may not tell enough about the future (causal) relationship between climate *change* and the change in economic outcomes. Another, not less challenging issue is that even if we correlate long-term temporal differences in weather with long-term differences in economic outcomes, some method is needed to disentangle the fraction of the long-term change in weather differences that is anthropogenic. This would offer useful information to calibrate the needed reduction in climate-altering emissions, let alone that regions need to be equipped with sufficient resilience capabilities to any weather extreme, regardless of its origin.

5.3 Spatial Ricardian Models

Spatial climate econometrics emerged as an extension of the Ricardian approach to climate impact assessment (see Auffhammer 2018, pp. 41–43; De Salvo et al. 2014). The Ricardian approach was pioneered by Mendelsohn et al. (1994) as a method to infer the willingness to pay in agriculture to keep the climate stable. The Ricardian approach improved upon previous "naive farmer scenarios" that assumed invariance in agricultural practice with respect to changing climate indicators.

Following De Salvo et al. (2014), a Ricardian econometric model is deduced from the solution to the following farmers' inter-temporal optimization problem:

$$V = \int P_{LE} e^{-\phi t} \tag{5.1}$$

where V is the discounted present value of net revenues per hectare P_{LE}, using the discount rate ϕ (t denotes time). Net revenues are equal to

$$P_{LE} = \sum_i P_i Q_i(X, F, Z) - \sum_i RX \tag{5.2}$$

where P_i and Q_i are, respectively, the market price and output of crop i; F is a vector of climate variables; Z is a vector of soil and economic variables; X is a vector of purchased inputs (excluding land) and R is a vector of input prices. Agricultural producers are assumed to operate in perfectly competitive markets for both inputs and outputs. Interest rates, rates of capital gains and capital per acre are assumed equal for all plots of land, ensuring proportionality between land value and land rent.

Solving the above optimization problem yields a relationship between the value of agricultural outputs and climate variables, controlling for soil and economic conditions. The sign and magnitude of such relationship implicitly discounts adaptation, as the theoretical model in Eq. 5.2 implies that farmers are ready to switch crops if this can lead to higher profits. This is considered a key advantage of the Ricardian approach. Depending on the function $Q_i(X, F, Z)$, the econometric specification can involve a linear or non-linear impact of climate on the farmer's outputs of interest. Many applications include polynomials of climate variables on the right-hand side to account for non-linearities.

De Salvo et al. (2014) provide a detailed overview of the many applications of the Ricardian approach, partly updated by Auffhammer (2018). One of the main methodological limitations highlighted by De Salvo et al. (2014) concerns the estimation bias and inefficiency attributable to spatial correlation. As mentioned before, if information could be costlessly transferred across space, the influence of neighbours on a location would not matter and spatial effects in climate impacts could be ignored. The original Ricardian model would still work. Instead, the existence of a tacit component in information and knowledge, whose transmission requires spatial proximity, makes the case for a new rendition of the Ricardian model wherein distance from other farms would be impactful on a farm's decisions.

The spatial Ricardian model, in a way, provides an appropriate solution to the omitted variables problem that hampers the credibility of cross-sectional and panel data estimates of climate impacts. As regards the relevance of space, both cross-sectional and panel regressions suffer from estimation issues. Howard and Sterner (2017) point out that cross-section estimates cannot take care of variables that display little spatial variability. With the advantage of the time dimension, panel regression mitigates the omitted variables bias through fixed effects by location, while exploiting a longer observational window. Still, in case of measurement error in weather indicators, including more fixed effects can amplify measurement error biases.

Needles to say, estimating a spatial version of the Ricardian model is only legitimate if so is suggested by the tests for spatial effects.[4] Which specification of the spatial Ricardian model is more appropriate is still debatable. Model specification requires answering to two sets of questions. First, which variables are spatially autocorrelated: the outcome variable, the error term, the explanatory variables (i.e. weather indicators) or combinations thereof? This is preliminary to the second question, about measures of spatial interactions. The modelling choices related to both questions are constrained by data availability, and, in particular, by the spatial granularity of the data.

As it will be clear from the literature review, most works in this area have adopted spatial econometric specifications with the primary goal of dealing with spatial autocorrelation in land value. In policy terms, this is not irrelevant, as it helps gaining insights on physical spillovers and on the spatial transmission of information and knowledge about adaptation practices. As investigated in the literature on innovation clusters (e.g. in the book by Maffioli et al. 2016) and in programme evaluation with spillover effects (Angelucci and Di Maro 2015), the existence of positive spatial autocorrelation in the outcome variable of interest implies that the causal effect of a public stimulus on the system is magnified by spatial propagation. Therefore, any desired effect of a policy can be attained through a smaller "dose" of public money. Accordingly, a spatial autocorrelation model should be estimated, and the next modelling decision would involve the spatial-weighting matrix associated to the land values of the neighbouring units (farms, counties, regions, depending on the spatial granularity of the data). It is common practice, though, to test the SAR model against a SEM model, since spatial autocorrelation may concern unobserved determinants of land value or otherwise defined exogenous disturbances. However, the explanatory variable of interest may not be a policy measure, but rather a variable that is outside of the immediate control of either farmers or policy-makers (such as climate). In such a case, one wonders if confining spatial effects to just the outcome would not distort information on the causal mechanism at work.

[4] See Chaps. 3 and 4 for theoretical and practical indications on statistical tests for spatial effects.

5.4 Estimates of Spatial Climate Spillovers on Land Value

Within the flourishing Ricardian literature, countless papers have extended the original framework to tackle the problem of spatial correlation. Spatial correlation has been accounted for by including additional explanatory variables referring to a larger spatial scale (Polsky and Easterling 2001), by means of non-parametric corrections for the standard errors of the estimated models (Deschênes and Greenstone 2007; Schlenker and Roberts 2009) and through spatially autocorrelated omitted variables (as in Schlenker et al. 2005, 2006).[5]

Hereby, we prefer to illustrate in a somewhat deeper fashion the contributions that more explicitly manage to estimate spillover effects. We believe this to be an essential viewpoint on climate impact assessment, as it provides a rigorous quantification of indirect damages or benefits, responding to Burke et al. 2016 call (see Sect. 5.1). To our knowledge, in this respect, the most exhaustive work to date is Dall'Erba and Domínguez (2016), who cite Polsky (2004), Seo (2008) as the only previous works that explicitly estimated spillover effects.

Polsky (2004) estimated econometric models for the per acre agricultural land value (measured in 1992 US dollars) including spatial lags and a Groupwise Heteroscedastic (GHET) term, on cross-sections of county-level data for the Great Plains (USA). Each cross-section referred to a different year (1969, 1974, 1978, 1982, 1987, 1992). Non-zero entries in the spatially weighted matrix capture the existence of a border between counties, a simplistic way of taking account of social interactions among farmers in spatially proximate land, whereas the GHET term was included with the goal of reducing the possible omitted variables bias and improving the efficiency of the estimators. Weather variables comprise the mean accumulated monthly amount of precipitations and the number of growing degree days for selected months. The analysis took account of the fact that counties belong to two different subregions of the Great Plains, divided by the Ogalalla Aquifer. Such a geographical division creates further omitted variable issues that the authors tackle. The results confirmed the initial insight about spatial autocorrelation in land values, presumably based on spatial proximity making communication of new farming techniques easier, net of similarity in soil characteristics across borders.

In a subsequent work by Seo (2008), the adoption of a spatial econometric approach is motivated by the concern over a possible misinterpretation of Ricardian model results. In particular, the author argues that differences in land values among farms or counties, estimated through non-spatial regression models, could be mistakenly understood as due to climate variation. In order to provide unbiased estimates, spatial lag and spatial error models of land value are estimated on farm-level data from South American countries (Argentine, Uruguay, Chile, Brazil, Venezuela, Ecuador, and Colombia) observed between July 2003 and June 2004. Farm-level data are collected through interviews in selected districts. Two alternative spatial-weighting matrices are considered, including district land values and provincial land values. Weather variables of interest include average temperatures and precipita-

[5] See Auffhammer 2018 and De Salvo et al. 2014 for more references.

tions in different seasons. Findings highlight that estimates of climate damages after accounting for spatial autocorrelation are much smaller, up to an order of magnitude smaller than those from non-spatial specifications (such as OLS and fixed effect panel models).

The paper by Dall'Erba and Domínguez (2016) on counties in Southwestern United States (Colorado, New Mexico, Arizona, Nevada) improves along a number of dimensions while confirming some previous modelling choices. On the innovative side, it is worth mentioning that first, the paper contains an in-depth discussion on the choice of the weather variables. Much attention is put on avoiding collinearity (e.g. by not including squares) and on properly measuring extreme weather events, something missing in Polsky (2004), Seo (2008). Weather variables include average summer temperatures, average winter and summer precipitation, maximum daily precipitation for each season, plus new variables that account for extreme heat and cold events.[6]

As another novelty, the authors take an agnostic standpoint on the modelling side and, perhaps first in the literature, consider the possibility of climate spillovers. This is a significant innovation, as previous works implicitly assumed that a climate shock would affect land value in other locations only through unobserved factors or inter-farmers' communication channels that lead to adaptation in agricultural practices. Most importantly, Dall'Erba and Domínguez (2016) mention several instances of how a climate shock in one location can determine changes in climate elsewhere, due to interconnections among natural capital stocks.[7] Accordingly, the authors estimate a spatial lag model and compare it to a spatial error model that includes also spatial autocorrelation in climate variables. The advantage of the latter, with respect to non-spatial specifications of the Ricardian model, is illustrated in the possibility to decompose the marginal effect on land value in a direct effect and a spillover effect, namely, the effect of weather event experienced in the immediate, first-order, neighbouring counties.

Confirming previous insights, the spatially weighted matrix is framed in a way that proxies the relative strength of different counties in influencing each other through the diffusion of farming techniques. In particular, the (per acre) agricultural output ratios between counties are included as entries of the W matrix (assumed common for both the climate variables and the error term).

---

[6]Specifically, the author has computed a moving window of temperatures for 15 days (7 before and 7 after) and counted the number of events above the upper (or lower) 90th percentile of the entire temperature distribution.

[7]Examples include intense precipitation due to summer thunderstorms leads to floods, property damages and casualties across several nearby counties; irrigated water that is crucial to Arizona's and New Mexico's agriculture originates hundreds of miles away in the Colorado Rockies; public spending in agricultural R&D can spill over neighbouring states.

5.5 Outlook and Conclusion

In concluding this detour on the role of space in climate impact assessment for agricultural values, it is worth adding a few considerations about the methodological caveats and challenges ahead. While some econometric issues have already been touched upon in the previous sections, what needs to be carefully investigated in future research is the status of econometric techniques and results vis-á-vis policy-making expectations and requirements. This involves, in the first place, specific research on the informational content of spatial econometric estimates for adaptation and mitigation policies. Not less importantly, econometric approaches to climate impact assessment compete—but may be combined—with alternative approaches, grounded on simulation techniques, that at the same time happen to paint more persuasive pictures of the future dangers, while possessing a longer pedigree in influencing policy and stakeholders, as through institutions such as the IPCC and COP meetings. We will deal with both issues in turn, in the following subsections.

5.5.1 Resilience, Adaptation and Mitigation

The coefficients that capture the average response of economic actors to weather events constitute econometric estimates of damage functions. Depending on the econometric specification, coefficients provide information on the "mechanical" effect of weather on the outcome of interest, net of individual responses driven by profit and utility considerations, commonly termed as adaptation.

In relation to climate change, resilience is commonly understood to embrace both the capacity to withstand the disruptive effects of climate change and the speed and flexibility of response and recovery. As such, regions can improve their resilience if appropriate mitigation and adaptation activities are put in place. More specifically, adaptation can increase the speed of recovery, whereas mitigation can make the system more robust to shocks (McDaniels et al. 2008).

Auffhammer (2018) makes a number of relevant observations for the design and interpretation of econometric resilience assessments in the context of global warming. One is that in most economic sectors, the long run allows for a wider array of adaptation options. Economic agents in the long run can change their habits, technologies, input combinations, land use and, quite relevantly for spatial econometrics, their location. Shorter term resilience assessments would, therefore, overstate the negative impacts of climate change. In fact, in an evolutionary perspective on economic adaptation, one may as well argue that actual (short-term) damages act as powerful motivators for the adoption of behaviours or technologies that allow to reduce impacts in the long run.[8] Another insightful remark by Auffhammer (2018),

[8]This is not to say that the threat of damages is not an incentive for adaptation, but uncertainty on the magnitude of damages and cognitive psychology effects, such as saliency effects, may weaken such incentives.

based on an example about water withdrawals by farmers, is that short-run adaptation can give rise to negative environmental externalities in the long run, such as resource depletion. In that case, econometric estimates showing a region's ability to withstand climate shocks in the short run may not be a good news as it appears.

Critically, the Ricardian approach assumes perfect competition and leaves the public sector in the background. In other words, the assumption is that only adaptation by self-seeking individuals exists, and hence Ricardian coefficients incorporate information on autonomous adaptation, but not on planned adaptation guided by public actors. However, a farmer's ability to cope with shocks might depend on public programmes that improve all agent's resilience, albeit to possibly different extents, or even on positive externalities from other agents' adaptive efforts. The inclusion of proxies for public programmes or other contextual and institutional variables in Ricardian models is not widespread. But besides, it is arguable that spatial dependence may affect the very coefficients that measure the impact of climate on land value. One may distinguish between a gross response, measuring the climate-outcome nexus in case of no adaptation, and a net response, which is obtained by subtracting the offsetting effect of adaptation. The difference between gross and net response, though, may only in part be attributable to the agent's own adaptation strategy. It may reflect her neighbours' adaptation as well as planned adaptation strategies put in place by public bodies.

5.5.2 Econometric and Simulative Impact Assessment: From Competition to Cooperation

In the perspective of anticipatory adaptation, private and public decision-makers need projections on changes in the frequency and intensity of weather events, and on the associated vulnerabilities of sectors, regions and societal groups. Prediction and assessment methods in climate economics, notably those adopted by the IPCC, rely on various numerical methodologies that have been used to estimate the potential damages from climate change.

Integrated assessment models (also known as IAMs) and computable General Equilibrium Models (CGE) are the mainstream families of models for the joint assessment of economic and climate dynamics. IAMs, such as the Dynamic Integrated Climate-Economy (DICE) (Nordhaus 1994), the Framework for Uncertainty, Negotiation and Distribution (FUND) (Tol et al. 1995) and Policy Analysis of the Greenhouse Effect (PAGE) (Plambeck and Hope 1996), feed socio-economic scenarios, that allow to compute future emissions trajectories, into a climate model that translates emissions paths into temperature changes. Damage functions map temperature anomalies or other climate outcomes into economic damages. CGE models, such as the Inter-temporal Computable Equilibrium System (ICES) (Bosello et al. 2012) and Environmental Impact and Sustainability Applied General Equilibrium Model (ENVISAGE) (Roson and Van der Mensbrugghe 2012), managed to account for general equilibrium effects that were only imperfectly captured by IAMs. Both IAMs and CGEs fail to account for catastrophic climate change, non-market damages and the cost of the transition to sustainable production frameworks. Both model families are based on the representative-agent assumption, and hence they cannot be truly useful as regards the assessment of spatial impacts and interactions.

IAMs have, moreover, been criticized, among other reasons, because of ad hoc assumptions on the damage function (Pindyck 2013). For example, Nordhaus (2008) uses an inverse quadratic loss function, Weitzman (2009) proposes a negative exponential functional specification emphasizing the catastrophic role of large climate changes. In this respect, IAMs are not less debatable than econometric models. In both approaches, the functional form of the damage function is assumed, e.g. based on parsimony considerations; sometimes, it is deduced from microeconomic axioms. Econometric models attempt to learn about the coefficients from real-world data that, however, can only provide a retrospective and static view on the climate-economy nexus. Simulation models follow a scenario approach, i.e. modellers in that approach run simulations under various sets of parameter values (not necessarily based on real-world estimates).

As a most problematic issue from the viewpoint of spatial resilience, the economic damage is often modelled as a fractional GDP loss, interpreted as global willingness to pay to avoid the adverse impacts of climate change (e.g. in Howard and Sterner 2017). A GDP loss is—after all—nothing but the sum of microeconomic losses occurring to different firms and workers in different sectors and geographical areas. As a matter of fact, what really matters for adaptation, mitigation and resilience is an assessment of how different microeconomic units are affected and how damages propagate through space. Propagation patterns can markedly differ even if they give rise to the same fractional GDP loss. Accordingly, Hsiang and Kopp (2018) in their study on the distribution of climate damages highlight that the social cost from climate shocks is made of two components: the level of exposure to environmental conditions and a vector of socio-economic attributes that may influence how exposure affects measures of economic well-being. This is consistent with an earlier call by Schröter et al. 2005 for a "place-based" assessment of vulnerability, performed at a spatial scale that corresponds to the scale of decision-making by policy-makers and stakeholders, without, however, underestimating that governance at different aggregation levels matters for climate policy decisions.

A bottom-up approach that aims to overcome this limitation has been proposed by Lamperti et al. (2018) in their DSK model. Casting itself in the complexity approach (e.g. Gallegati and Kirman 2012; Balint et al. 2017; Dosi and Roventini 2019), the DSK model eschews the assumption of a representative agent and predicts the lack of isomorphism between the micro- and macroeconomic outcomes as an emerging property of the economic system dynamics. In the DSK model, the damage generating function is stochastic and evolves over time according to climate dynamics. The size of the shock affecting economic agents is a draw from a probability distribution, whose parameters depend on the temperature anomaly.[9] "Unpacking" the aggregate damages into microeconomic ones allows to investigate how the effects of micro-climate shocks are amplified by the interactions of heterogeneous agents along an evolving network structure of business relationships, even though the DSK model is not explicitly spatial. A similar approach is diffusing in the agent-based literature (see also Dafermos et al. 2017; Lamperti et al. 2019, 2020).

Simulation models equipped with modules that allow for a fine microeconomic and spatial assessment of the direct and indirect effects of climate shocks could supply extremely valuable inputs and improve climate econometric models. Available projections can be useful in various respects. They can help identifying the most salient weather extremes and the most exposed locations, allowing to refine the process of sample selection. Projections can also be included as explanatory variables in spatial econometric models, as they can serve as proxies for expected climate, freeing the econometric approach from over-reliance on retrospective measures. Severen et al. 2018 provide an interesting extension of the Ricardian method by incorporating climate expectations. So far, regression coefficient estimates have been used in order to predict how the outcome variables may change under different scenarios, supplied by IPCC or other organizations, which in turn rely on IAM or CGE analysis. Clearly, nothing guarantees that econometric coefficients will remain stable, nor that econometric models have sufficient external validity as to inform policy-makers on potential damages in the future or in other (however similar) locations.

A specific issue for spatial econometrics concerns the ability of simulating future changes in the spatially weighted matrix. Climate change may indeed alter the existing strength of spatial linkages. This is most evident with respect to works using output measures as entries in the spatially weighted matrix (such as Seo 2008 or Dall'Erba and Domínguez 2016): in that case, the spatially weighted matrix is endogenous to climate on a long time horizon. Not less relevant is the possible influence of climate change on other types of spatially weighted matrices. Information spreading channels can be severed by climate-induced political upheaval, disruption of transportation infrastructures due to natural disasters and so forth.

Conversely, agent-based IAMs are challenged to prove their ability to reproduce spatial patterns in the climate-economy nexus that are highlighted in the reviewed literature. As illustrated by Haldane and Turrell (2018), agent-based models prove their explanatory power by means of validation techniques applied to a wide array of stylised facts, at the micro- and macroeconomic levels (Windrum et al. 2007; Fagiolo et al. 2019). The facts against which ABMs have been validated so far are based on time-series and panel regression models. Validation against spatial econometric evidence is a promising research avenue that may cross-fertilize the fields of computational economics and econometrics.

---

[9]Following Ackerman et al. (2010), Weitzman (2011), Pindyck (2012), a suitable probability distribution for climate damages should allow for skewness and fat tails, as the probability of very large damages is higher than in a Gaussian process.

References

Ackerman, F., Stanton, E. A., & Bueno, R. (2010). Fat tails, exponents, extreme uncertainty: Simulating catastrophe in DICE. *Ecological Economics, 69*(8), 1657–1665.

Angelucci, M., & Di Maro, V. (2015). *Program evaluation and spillover effects*. The World Bank.

Auffhammer, M. (2018). Quantifying economic damages from climate change. *Journal of Economic Perspectives, 32*(4), 33–52.

Balint, T., Lamperti, F., Mandel, A., Napoletano, M., Roventini, A., & Sapio, A. (2017). Complexity and the economics of climate change: A survey and a look forward. *Ecological Economics, 138*, 252–265.

Behringer, W. (2010). A cultural history of climate. *Polity*.

Burke, M., Craxton, M., Kolstad, C. D., & Onda, C. (2016). Some research challenges in the economics of climate change. *Climate Change Economics, 7*(02), 1650002.

Carleton, T. A., & Hsiang, S. M. (2016). Social and economic impacts of climate. *Science, 353*(6304):aad9837.

Dafermos, Y., Nikolaidi, M., & Galanis, G. (2017). A stock-ow-fund ecological macroeconomic model. *Ecological Economics, 131*, 191–207.

Dall'Erba, S., & Domínguez, F. (2016). The impact of climate change on agriculture in the Southwestern United States: The Ricardian approach revisited. *Spatial Economic Analysis, 11*(1), 46–66.

Dell, Melissa, Jones, Benjamin F., & Olken, Benjamin A. (2014). What do we learn from the weather? The new climate economy literature. *Journal of Economic Literature, 52*(3), 740–798.

De Salvo, M., Begalli, D., & Signorello, G. (2014). The Ricardian analysis twenty years after the original model: Evolution, unresolved issues and empirical problems. *Journal of Development and Agricultural Economics, 6*(3), 124–131.

Deschênes, O., & Greenstone, M. (2007). The economic impacts of climate change: evidence from agricultural output and random fluctuations in weather. *American Economic Review, 97*(1), 354–385.

Dosi, G., & Roventini, A. (2019). More is different... and complex! the case for agent-based macroeconomics. *Journal of Evolutionary Economics, 29*(1), 1–37.

Fagiolo, G., Guerini, M., Lamperti, F., Moneta, A., & Roventini, A. (2019). Validation of agent-based models in economics and finance. In *Computer simulation validation* (pp. 763–787). Cham: Springer.

Fingleton, B. (2018). Exploring Brexit with dynamic spatial panel models: Some possible outcomes for employment across the EU regions. *The Annals of Regional Science* 1–37.

Fingleton, B. (2019). Italexit, is it another Brexit? *Journal of Geographical Systems* 1–28.

Gallegati, M., & Kirman, A. (2012). Reconstructing economics: Agent based models and complexity. *Complexity Economics, 1*(1), 5–31.

Haldane, A. G., & Turrell, A. E. (2018). An interdisciplinary model for macroeconomics. *Oxford Review of Economic Policy, 34*(1–2), 219–251.

Howard, P. H., & Sterner, T. (2017). Few and not so far between: A meta-analysis of climate damage estimates. *Environmental and Resource Economics, 68*(1), 197–225.

Hsiang, S. (2016). Climate econometrics. *Annual Review of Resource Economics, 8*, 43–75.

Hsiang, S., & Kopp, R. E. (2018). An economist's guide to climate change science. *Journal of Economic Perspectives, 32*(4), 3–32.

Hsiang, S., Oliva, P., & Walker, R. (2019). The distribution of environmental damages. *Review of Environmental Economics and Policy, 13*(1), 83–103.

Katz, R. W., & Brown, B. G. (1992). Extreme events in a changing climate: Variability is more important than averages. *Climatic Change, 21*(3), 289–302.

Lamperti, F., Dosi, G., Napoletano, M., Roventini, A., & Sapio, A. (2018). Faraway, so close: Coupled climate and economic dynamics in an agent-based integrated assessment model. *Ecological Economics, 150*, 315–339.

Lamperti, F., Mandel, A., Napoletano, M., Sapio, A., Roventini, A., Balint, T., et al. (2019). Towards agent-based integrated assessment models: Examples, challenges, and future developments. *Regional Environmental Change, 19*(3), 747–762.

Lamperti, F., Dosi, G., Napoletano, M., Roventini, A., & Sapio, A. (2020). And then he wasn't a she: Climate change and green transitions in an agent-based integrated assessment model. *Forthcoming, Technological Forecasting & Social Change.*

Maffioli, A., Pietrobelli, C., & Stucchi, R. (Eds.). (2016). The impact evaluation of cluster development programs: Methods and practices. Inter-American Development Bank.

McDaniels, T., Chang, S., Cole, D., Mikawoz, J., & Longstaff, H. (2008). Fostering resilience to extreme events within infrastructure systems: Characterizing decision contexts for mitigation and adaptation. *Global Environmental Change, 18*(2), 310–318.

Mendelsohn, R., Nordhaus, W. D., & Shaw, D. (1994). The impact of global warming on agriculture: A Ricardian analysis. *The American Economic Review, 84*(4), 753–771.

Nordhaus, W. D. (2008). *A question of balance: Economic modeling of global warming*. New Haven: Yale University Press.

Pindyck, R. S. (2012). Uncertain outcomes and climate change policy. *Journal of Environmental Economics and Management, 63*(3), 289–303.

Pindyck, R. S. (2013). Climate change policy: What do the models tell us? *Journal of Economic Literature, 51*(3), 860–872.

Plambeck, E. L., & Hope, C. (1996). PAGE95: An updated valuation of the impacts of global warming. *Energy Policy, 24*(9), 783–793.

Polanyi, M. (1966). *The tacit dimension*. University of Chicago press.

Polsky, C. (2004). Putting space and time in Ricardian climate change impact studies: agriculture in the US Great Plains, 1969–1992. *Annals of the Association of American Geographers, 94*(3), 549–564.

Polsky, C., & Easterling, W. E. (2001). Adaptation to climate variability and change in the US Great Plains: A multi-scale analysis of Ricardian climate sensitivities. *Agriculture, Ecosystems & Environment, 85*, 133–144.

Renton, M., Shackelford, N., & Standish, R. J. (2014). How will climate variability interact with long-term climate change to affect the persistence of plant species in fragmented landscapes? *Environmental Conservation, 41*, 110–121.

Revesz, R. L., Howard, P. H., Arrow, K., Goulder, L. H., Kopp, R. E., Livermore, M. A., et al. (2014). Global warming: Improve economic models of climate change. *Nature, 508*(7495), 173–175.

Romer, C. D. (1999). Changes in business cycles: Evidence and explanations. *Journal of Economic perspectives, 13*(2), 23–44.

Roson, R., & Van der Mensbrugghe, D. (2012). Climate change and economic growth: impacts and interactions. *International Journal of Sustainable Economy, 4*(3), 270–285.

Severen, C., Costello, C., & Deschenes, O. (2018). A Forward-Looking Ricardian Approach: Do land markets capitalize climate change forecasts?. *Journal of Environmental Economics and Management, 89*, 235–254.

Seo, S. N. (2008). Assessing relative performance of econometric models in measuring the impact of climate change on agriculture using spatial autoregression. *Review of Regional Studies, 38*(2), 195–209.

Schlenker, W., & Roberts, M. J. (2009). Nonlinear temperature effects indicate severe damages to US crop yields under climate change. *Proceedings of the National Academy of sciences, 106*(37), 15594–15598.

Schlenker, W., Hanemann, W. M., & Fisher, A. C. (2005). Will US agriculture really benefit from global warming? Accounting for irrigation in the hedonic approach. *American Economic Review, 95*(1), 395–406.

Schlenker, W., Hanemann, M., & Fisher, A. (2006). The impact of global warming on U.S. agriculture: An econometric analysis of optimal growing conditions. *Review of Economics and Statistics, 88*, 113–125.

Schröter, D., Polsky, C., & Patt, A. G. (2005). Assessing vulnerabilities to the effects of global change: an eight step approach. *Mitigation and Adaptation Strategies for Global Change, 10*(4), 573–595.

Thomalla, F., Downing, T., Spanger-Siegfried, E., Han, G., & Rockström, J. (2006). Reducing hazard vulnerability: Towards a common approach between disaster risk reduction and climate adaptation. *Disasters, 30*(1), 39–48.

Tol, R. S. J., Van Der Burg, T., Jansen, H. M. A., & Verbruggen, H. (1995). Socio-economic aspects of the greenhouse effect: Climate fund. In *Studies in Environmental Science* (Vol. 65, pp. 1283–1288). Elsevier.

Weitzman, M. L. (2009). On modeling and interpreting the economics of catastrophic climate change. *The Review of Economics and Statistics, 91*(1), 1–19.

Weitzman, M. L. (2011). Fat-tailed uncertainty in the economics of catastrophic climate change. *Review of Environmental Economics and Policy, 5*(2), 275–292.

Windrum, P., Fagiolo, G., & Moneta, A. (2007). Empirical validation of agent-based models: Alternatives and prospects. *Journal of Artificial Societies and Social Simulation, 10*(2), 8.

Chapter 6
Resilience to Climate Impacts and Spatial Propagation in the Power Industry

Abstract This chapter reviews applications of spatial econometrics that are relevant for understanding the resilience of the electricity industry to climate-related impacts. Two domains of spatial econometric applications are considered, namely, wholesale power demand and the residential uptake of photovoltaic panels. The results so far published on power demand are interpreted in light of a novel theoretical framework, wherein the resistance of the electricity market is parametrized on the spatial correlation coefficient. The literature review on photovoltaic panel adoption is preceded by the illustration of the theoretical insights behind the role of space in technology diffusion. Implications for the resilience of the industry to shocks are drawn.

Keywords Resilience · Resistance · Power market · Spatial econometrics · Electricity demand · Technology adoption · Photovoltaic panels

6.1 Introduction

The power industry plays a dual role in the current debate on climate policy tools and objectives. On the one hand, the power industry is among the major contributors to climate-altering emissions, due to fossil fuelled plants that still represent the standard power generation technology in most countries in the world. Reliance on coal, natural gas and oil is the heritage of the centralized power generation paradigm that became dominant in the early twentieth century (Granovetter and McGuire 1998). Vested interests hamper the energy transition in resource-rich countries.

On the other hand, the power industry enjoys wide opportunities for decarbonization, thanks to the burgeoning technological progress in renewable energy and the resurgence of the decentralized power generation paradigm. Public programmes have been essential in this process, whether in terms of market deregulation (starting with the 1978 PURPA in the United States) or through subsidies and tax cuts (to meet the targets set since the Electricity Production from Renewable Energy Sources Directive 2001/77/EC and Renewable Energy Directive 2009/28/EC). Therefore, when it comes to climate change, the power industry is both a problem and a solution (Helm and Hepburn 2019).

R. De Siano et al., *Regional Resilience to Climate and Environmental Shocks*, SpringerBriefs in Regional Science, https://doi.org/10.1007/978-3-030-54588-8_6

The physical assets and infrastructures that constitute the power industry are themselves exposed to climate-induced shocks. Several instances of climate-related effects on the power industry have been studied in the literature (Schaeffer et al. 2012, Panteli and Mancarella 2015, Chandramowli and Felder 2014). Bonjean Stanton et al. (2016) offer a systematic review of the impacts of climate change and variability on power systems, including applications of econometric and simulation methods. The closest to layman's experience is the change in energy use habits due to unusually warm or cold seasons. Spatial patterns of energy demand, though, affect the use of transmission infrastructures, making some regions more frequently congested out, or others more integrated and therefore more exposed to volatility sparked by events elsewhere in the network. Higher temperatures, moreover, hamper the transport capability of transmission lines and increase the energy losses. Overhead lines and towers can be damaged by or default because of windstorms, cold waves and heavy snow. Concerning power generation, climate change affects both resource availability and the supply of electricity. For instance, increasing wind speeds can lead to energy curtailments. Rising sea levels can threaten oil and gas pipelines, whereas changing groundwater levels affect coal quality and handling, thereby altering the supply of fossil fuels for power generating units. On the other hand, more ice-free summers in the Arctic Region can make the drilling seasons longer. Droughts and reduced snowmelt affect the availability of water for hydropower generation and for cooling of thermal and nuclear power plants. Changing patterns of insolation may cause the existing investments in photovoltaic and concentrated solar power facilities to run into stranded asset problems. Consequences may range from erratic variation and curtailments in power generation to reduced equipment safety and lifetime.

The exposure of the power industry to climate impacts bears potentially more disruptive consequences than most other facilities. Indeed, power generation and transmission facilities can be considered critical infrastructures, namely, infrastructures that are fundamental for the functioning of the economy (Künneke et al. 2010). Such infrastructures house assets and perform functions that are critical in order to meet expectations of the users, such as reliability, safety, and security of supply.[1] Furthermore, but not less importantly, electricity is a *General-Purpose Technology* (GPT), namely, a technology that is used in all sectors of the economy (pervasiveness) and that holds inherent potential for technical improvements and innovation complementarities (Bresnahan and Trajtenberg 1995). Shocks that make electricity more costly determine financial losses in the downstream sectors that use electricity as a production input.

The above considerations imply that the exposure of the power industry to climate shocks deserves to be carefully analysed and monitored. Given the increasing

[1]In the power industry, reliability refers to the ability of making electricity available at all times to all users; safety concerns the ability to protect users from adverse consequences of technical malfunctioning, e.g. health damages due to blackouts, security of supply requires that primary energy sources be available at reasonable costs.

frequency and intensity of climate shocks and the efforts to mitigate them, intellectual efforts need to converge on adaptation strategies that make our critical infrastructures resilient.

In the power industry context, resilience is defined, following Panteli and Mancarella (2015), as "the ability of a power system to withstand extraordinary and high impact-low probability events such as due to extreme weather, rapidly recover from such disruptive events and absorb lessons for adapting its operation and structure to prevent or mitigate the impact of similar events in the future". In their analysis, the authors interpret the definition from an operational viewpoint: resilience is related to the speed with which the weather-damaged components are restored. In an economic perspective, and on a longer time horizon, resilience can alternatively be seen as the ability of the system to reallocate its power demand and generating assets in ways that restore the pre-change cost of electricity. More specifically, in this chapter we explore the resilience of the power market in terms of resistance, defined as the ability of the market to resist the initial impact of the crisis (see Chap. 2 for an analogous definition concerning regions). This approach measures resistance of the market with respect to a benchmark position, measuring the trend of the market in case the change had not occurred.

This chapter acknowledges that in understanding the exposure and resilience of the power industry to climate impacts, the spatial patterns in propagation of the effects and in the diffusion of adaptation capabilities cannot be overlooked.

Thanks to the pathbreaking insights by Nijkamp (1983) on the spatial analysis of energy production and the theoretical work by Bohn et al. (1984) on nodal pricing, the spatial organization of the electricity industry has entered the energy policy debate. Its relevance has been fostered by the goal of decarbonization. Balta-Ozkan et al. (2015a) identify five trends in empirical evidence and in economic thinking that motivate the study of the electricity industry through spatial analysis.[2]

First, potentials for renewable energy generation are spatially uneven due to geographical differences in average insolation rates, wind speeds, and geothermal and hydrological endowments, which reflect local climatic and orographic conditions; leading to a second trend, the deployment of green energy technologies is spatially concentrated. Third, geographical fragmentation in the governance of liberalized electricity markets implies that relationships between electricity users and producers unfold in space through both physical and social connections. Prosumers using distributed generation facilities are linked through social ties and their physical interlinkages are primarily local, as they are mediated through a (short-distance and low-voltage) distribution grid rather than through a (long-distance and high-voltage) transmission grid. Fourth and relatedly, geographical differences in urbanization rates influence the diffusion of distributed generation technologies. Finally, agglomeration economies are among the efficiency gains available through spatial integration between energy markets, but may be hampered by transmission constraints.

[2]See also De Siano and Sapio (2020), Boffa and Sapio (2015).

As a bottom line, the damages from climate shocks to the power industry and the opportunities for mitigation and adaptation strategies (such as decarbonization) are significantly conditioned by the spatial organization of power generation, transport and consumption. Despite this, spatial econometric methods have not yet gained a status comparable to time-series analysis in applied research work (De Siano and Sapio 2020).

This chapter reviews applications of spatial econometrics that are relevant for understanding the resilience of the electricity system. After discussing model specification issues in Sect. 6.2, 6.3 overviews the existing results on spatial correlation in power demand, interpreted in light of a theoretical framework for the assessment of electricity market resilience. Section 6.4 reviews the evidence from spatial econometric estimates of photovoltaic (PV) panels adoption. Implications for the robustness of the industry to shocks are drawn in the concluding Sect. 6.5, along with an outlook for future research.

6.2 Econometric Specification Issues

Before reviewing the spatial econometric evidence on power systems, it is worth dealing with model specification issues. As we will see, the finest spatial granularity in the analysis of power demand can be achieved by means of data from wholesale power exchanges, implying that model specification is conditioned by the institutional setting of the market. The spatial analysis of technology adoption by households, instead, may face other types of constraints, such as limitations in data collection.

The spatial analysis of wholesale power demand needs to embody a basic principle: in deregulated power exchanges, valuation of the commodity is ultimately related to the management of a spatial network. Indeed, trading involves contracts for the delivery of electrical power at pre-determined dates. Though, "delivery" of such a peculiar commodity as electricity means that the transmission system operator, based on the demand and supply schedules submitted by the market participants, needs to optimize the allocation of power flows in the network, under a set of technical constraints and notably under the constraint that power be available to all users connected to the network at all times (reliability).

For the above reasons, wholesale electricity prices are determined in spatially identified locations. In the nodal pricing system, prices are computed at each node of the transmission grid based on the bids and offers submitted at that node and on import/export flows pointing to it. In the zonal pricing system, the transmission grid is segmented in zones, i.e. subsets of the grid where congestion is minimized. Demand and supply are balanced at the national level, allowing to determine a system marginal price, unless congestion occurs between zones. In that case, prices are determined for each zone in a way that solves the congestions. Although nodal pricing is shown to be optimal with respect to allocative efficiency, optimal nodal prices are seldom implemented in practice (Green 2007). Indeed, in most markets, the zonal resolution is the finest available (Italy, Germany, the NordPool), whereas the PJM (Pennsylvania-New Jersey-Maryland interconnection) is one noteworthy case of market adopting nodal pricing.

Applications of spatial econometrics to power exchanges, therefore, rely on longitudinal datasets with a pre-defined spatial dimension. Each variable (price, demand, supply from fossil fuels, supply from renewables, etc.) can be associated to a location—a node or a zone. Zones may or may not coincide with administrative regions.[3] In some cases (e.g. PJM), for privacy reasons the market only publishes zonal averages of nodal prices.

Building spatially weighted matrices is simplified by zonal aggregation, at the cost of losing the information in the high geographical resolution guaranteed by the nodal granularity.[4] This cost is minimized if price differentials within zones are small (Walton and Tabors 1996). When information is only available for averages across nodes, spatial dependencies may be blurred. In a zone of given size, if the spatial dependence in nodal prices dies "fast" enough within a zone, one should not expect any spatial dependence. Hence, a relatively good performance of spatial econometric models requires that spatial dependencies across nodes go beyond the zonal boundaries.

The choice of spatial granularity bears implications for the choice of the spatially weighted matrix. In applications to power exchanges, geographical adjacency (i.e. the existence of a border between regions or countries), as such, is neither necessary nor sufficient to expect a spatial effect. Two places can be interconnected without sharing a physical border (e.g. an island with the mainland), whereas borders do not guarantee the existence of a transmission link.[5]

The simplest spatially weighted matrix is an adjacency matrix, with entries equal to 1 if the localities are interconnected, 0 otherwise. However, dummies treat all interconnections uniformly, whereas a larger interconnection is supposed to imply a tighter correlation between prices across zones. Hence, one could use the capacity of the transmission lines linking the zones instead of simple dummies. To account for the fact that a transmission line can be congested, one can set to 0 the entries corresponding to congested lines, if the time granularity of the data is hourly, or the fraction of hours is without congestion, if instead the model is specified on average daily prices. Alternatively, one can use the actually available capacity, which however could suffer from endogeneity issues.[6] Such rigidities in the set up of spatially weighted matrices do not exist in the analysis of energy technology adoption. The availability of individual-level data on adoption allows for distance-based W matrices.

---

[3]For instance, in Italy the southern zone includes Calabria, Basilicata and Apulia.

[4]The following considerations concern spatial correlation in power demand, whereas the choice of spatially weighted matrix for climatic explanatory variables is discussed in Chap. 5.

[5]This is sometimes reflecting history or geopolitical conditions. For instance, Baltic States have inherited well-developed interconnections with the former Soviet Union, but are still insufficiently connected with their EU partners. Another case in point is Israel, discussed in Yasner (2012).

[6]These more sophisticated choices imply time-varying spatially weighted matrices.

There is still lack of consensus on the most appropriate spatial modelling strategy in electricity markets research. One may argue that prices and quantities at different localities are directly correlated because traders arbitrage away price differentials, supporting SAR. Just as meaningfully, prices in neighbouring localities may be influenced by unforeseen technical failures or other events simultaneously occurring in different locations of the grid. This would motivate the use of an SEM model.

6.3 Power Demand and the Resilience of Power Systems

People are usually shocked to learn that one day, due to climate change, the shape of their country may change irreversibly; their city and the surrounding coastlines may disappear under the rising sea level. Destruction of the natural capital is already occurring due to faster and ever more unpredictable windstorms and wildfires, which may increasingly affect residential and commercial buildings, as well as production facilities and infrastructures. From the viewpoint of the energy system, damaged buildings and factories and the progressive loss in urbanized areas translate into a spatial redistribution of energy demand. Gradual change in climate may as well alter the existing geographical patterns in electricity demand, by making some regions warmer and other cooler than they currently are. The patterns of electricity imports and exports among regions and countries, for a given spatial distribution of power supply, are going to change. Climate change is therefore going to drive the electricity price upwards or downwards, on a longer or shorter horizon, depending on its influence on power demand. Upward pressure on the wholesale electricity price would make energy bills more expensive for end users.

In this context, resilience can be measured in terms of resistance, based on comparing the actual change in the electricity market equilibrium due to a structural change in the market, and a counterfactual change in the price level. Some hints at another aspect of resilience, namely recoverability, can be attained by studying how costly and, therefore, how fast it would be to implement adaptation strategies that restore the pre-existing price level. The focus on price is motivated by the consequences of a structural change in demand: it may aggravate the cost burden on final energy users if the price soars, but it may as well diminish incentives for power generation and investment if the change exercises a downward pressure on price.

In studying power market resilience, we take an explicitly spatial approach. Intuitively, if power demand is spatially correlated, any local change in demand propagates its effects across regions, multiplying its ultimate impact. Resilience, therefore, is more costly in terms of resources and time required to restore the initial state of the system than in absence of spatial correlation.

Section 6.3.1 presents a formal analysis that translates the above insights into a model of the wholesale electricity price. The theoretical results shed light on a novel interpretation of the existing evidence on spatial correlation in power demand, reviewed in Sect. 6.3.2.

6.3.1 Resilience of a Power Market to Structural Change in Power Demand

Goal of this section is to build a theoretical framework on wholesale power demand with a twofold goal. First, we seek to understand whether spatial correlation hampers power market resilience. Second, we formulate, through our theoretical framework, a retrospective interpretation of the spatial econometric evidence on power demand (to be reviewed in Sect. 6.3.2). Our theoretical model allows to suggest how the spatial correlation coefficient, estimated in the literature, provides information on the amount of resources needed to restore the system's conditions in relation to a given structural change. The change may be explicitly measured as a change in a weather variable. Alternatively, it may be part of the error term along with other disturbances, as in most spatial econometric works on power demand.[7]

Consider an electricity market composed of N zones, $i = 1, \ldots, N$. In each zone i, energy users submit inelastic demand schedules D_i, whereas the supply from power generating companies is an increasing function of the zonal price, i.e. $S_i = f(p_i)$.[8] Let us assume that supply is linear:

$$S_i = \alpha_i + \beta_i p_i \tag{6.1}$$

where α_i and $\beta_i > 0$ are zone-specific parameters of the supply function and α_i, in particular, measures the amount of price-inelastic power supply. Since renewables have null marginal costs and are remunerated through tariffs in most countries, α_i can be seen as measuring the supply from renewables. The magnitude of the β_i coefficient depends on the marginal costs of the "dispatchable" power units, such as those using fossil fuels.

With sufficient transmission capacity, power flows equalize prices across zones, namely, $p_i = p, \forall i$. The equilibrium price is found by solving for the following condition:

$$\sum_i S_i = \sum_i D_i \tag{6.2}$$

namely,

$$\sum_i \alpha_i + p \sum_i \beta_i = \sum_i D_i \tag{6.3}$$

The solution reads

$$p^* = \frac{\sum_i (D_i - \alpha_i)}{\sum_i \beta_i} \tag{6.4}$$

---

[7]Hereby we focus on a change in power demand that is supposedly permanent, or at least a long-lasting one. This is the reason why we prefer to rely on the notion of structural change rather than on (temporary, short-termed) shocks. The structural change will be modelled as a permanent change in one of the model parameters.

[8]Time indexes are omitted for simplicity.

or, in matrix notation,

$$p^* = \frac{\mathbf{1}^T\mathbf{D} - \mathbf{1}^T\boldsymbol{\alpha}}{\mathbf{1}^T\boldsymbol{\beta}} \tag{6.5}$$

where $\mathbf{1}^T$ is the (transposed) sum vector, $\mathbf{D}$ the vector of zonal power demands, and $\boldsymbol{\alpha}$ and $\boldsymbol{\beta}$ the vectors of zonal supply function coefficients.

Suppose the data generating process of power demand is a spatial lag model:

$$\mathbf{D} = \rho\mathbf{W}\mathbf{D} + \epsilon \tag{6.6}$$

where ρ and $\mathbf{W}$ are, respectively, the spatial correlation coefficient and the spatial-weighting matrix and ϵ is the vector of error terms, with constant mean $\boldsymbol{\mu}$. Solving the spatial lag model for $\mathbf{D}$ yields the reduced-form expression

$$\mathbf{D} = (\mathbf{I} - \rho\mathbf{W})^{-1}\epsilon \tag{6.7}$$

with $\mathbf{I}$ denoting the identity matrix. This reduced form can be plugged into Eq. 6.5 to yield, after taking expectations $E[.]$,

$$E[p^*] = \frac{\mathbf{1}^T(\mathbf{I} - \rho\mathbf{W})^{-1}\boldsymbol{\mu} - \mathbf{1}^T\boldsymbol{\alpha}}{\mathbf{1}^T\boldsymbol{\beta}} \tag{6.8}$$

As argued at the beginning of this section, resilience of the power system can be measured in terms of resistance and the ability of the system to restore its initial conditions after a structural change.

Consider a positive structural change in demand, induced by unexpectedly high temperature, hits demand in zone i and permanently changes the power demand generating process. This can be formalized as $\Delta\mu_i > 0$, where Δ is the time-difference operator. An increase in the electricity price is expected, leading to more expensive power consumption. Alternatively, a negative structural change can damage some electricity-consuming facilities, e.g. some industrial machinery, leading to lower wholesale electricity prices to the detriment of power generating companies. Formally, $\Delta\mu_i < 0$.

In Chap. 2, resistance was described (in its numerator) as the difference between the change in the target variable (e.g. regional employment) after the structural change, and a counterfactual (such as national employment). In the theoretical framework we are building here, because we have assumed that no congestion occurs between zones/regions, it is more appropriate to compare the change in the target variable (the wholesale electricity price) with a temporal counterfactual.

Suppose the electricity price would have changed anyway, for instance, that renewable energy capacity follows a positive trend in at least some zone. Formally, $\Delta\boldsymbol{\alpha} > 0$. If so, the counterfactual is built by assuming that no other parameter in the model changes over time:

$$\Delta E[p^*|\Delta\mu = 0] = \frac{-\mathbf{1}^T \Delta\boldsymbol{\alpha}}{\mathbf{1}^T \boldsymbol{\beta}} \tag{6.9}$$

where the notation $\Delta E[p^*|\Delta\mu = 0]$ means: expected change in the equilibrium price, conditional upon a constant expected value of the error term.

In order to build a measure of resistance, $\Delta E[p^*|\Delta\mu = 0]$ is compared with $\Delta E[p^*|\Delta\mu \neq 0]$, which reads

$$\Delta E[p^*|\Delta\mu \neq 0] = \frac{\mathbf{1}^T (\mathbf{I} - \rho\mathbf{W})^{-1} \Delta\boldsymbol{\mu} - \mathbf{1}^T \Delta\boldsymbol{\alpha}}{\mathbf{1}^T \boldsymbol{\beta}} \tag{6.10}$$

The numerator in the resistance measure is the initial price level, $E[p^*]$. Overall, the resistance indicator reads

$$Resis = \frac{\Delta E[p^*|\Delta\mu \neq 0] - \Delta E[p^*|\Delta\mu = 0]}{E[p^*]} \tag{6.11}$$

or, more explicitly, after some algebra,

$$Resis = \frac{\mathbf{1}^T (\mathbf{I} - \rho\mathbf{W})^{-1} \Delta\boldsymbol{\mu}}{\mathbf{1}^T (\mathbf{I} - \rho\mathbf{W})^{-1} \boldsymbol{\mu} - \mathbf{1}^T \boldsymbol{\alpha}} \tag{6.12}$$

As shown by the above equation, resistance depends on the spatial correlation coefficient ρ. More specifically, the marginal effect of a change in μ_i on resistance is given by

$$\frac{\partial Resis}{\partial \Delta\mu_i} = \frac{\mathbf{1}^T (\mathbf{I} - \rho\mathbf{W})^{-1} \boldsymbol{e}_1}{\mathbf{1}^T (\mathbf{I} - \rho\mathbf{W})^{-1} \boldsymbol{\mu} - \mathbf{1}^T \boldsymbol{\alpha}} < 0 \tag{6.13}$$

and is negative. The intuition is that, due to higher spatial correlation, the effects of a structural change localized in a market zone are spread to other zones, therefore causing the electricity price to drift farther from the counterfactual.

To understand this point, consider two polar cases. On the one hand, suppose $\Delta\alpha > 0$, so that $E[p^*]$ decreases in the counterfactual. A negative change in demand ($\Delta\mu < 0$) reinforces this tendency, but the existence of positive spatial correlation makes price decline even more rapidly. On the other hand, suppose again $\Delta\alpha > 0$. A downward pressure on price would ensue, but if a positive structural change in demand hits the market ($\Delta\mu > 0$), this would offset it. In fact, because of positive spatial correlation, the electricity price may even increase above the pre-existing price level. Again, spatial correlation implies a widening difference between the actual price dynamics and the counterfactual.

Another way of exploring resilience to structural change in power demand involves studying the marginal effect on electricity prices and the amount of renewable energy capacity that needs to be built or scrapped in order to restore the pre-change state of the market. Indeed, if the supply function parameters $(\boldsymbol{\alpha}, \boldsymbol{\beta})$ remain constant, i.e. if supply from renewables and the costs of dispatchable sources do not change, the wholesale electricity price remains permanently higher or lower than before the change. The marginal effect of a change in μ_i is

$$\frac{\partial E[p^*]}{\partial \mu_i} = \frac{\mathbf{1}^T(\mathbf{I}-\rho\mathbf{W})^{-1}\boldsymbol{e}_1}{\mathbf{1}^T\boldsymbol{\beta}} > 0 \tag{6.14}$$

where $\boldsymbol{e}_1$ is the unit vector $(1, 0, \ldots, 0)$. Because ρ appears in the above formula, spatial correlation tunes the price sensitivity to structural change in demand. If $\rho = 0$, namely, if power demand in different zones are uncorrelated, then the impact of the change is simply the inverse of the sum of $\boldsymbol{\beta}$ coefficients:

$$\frac{\partial E[p^*|\rho = 0]}{\partial \mu_i} = \frac{1}{\mathbf{1}^T\boldsymbol{\beta}} \tag{6.15}$$

The power system is resilient if the parameters of the supply function change and sterilize the threat of a permanently higher (or lower) price level. This may occur endogenously or through policy; it may take a shorter or longer spell. For instance, if power supply from renewables grows in zone j, the price increase can be offset, consistent with the merit order effect detected in the literature.

To see this, consider the changes in μ_i and α_j that would keep the expected value of p constant. Those can be determined by setting to 0 the total differential, i.e. by solving

$$dE[p^*] = \frac{\partial E[p^*]}{\partial \mu_i} d\mu_i + \frac{\partial E[p^*]}{\partial \alpha_j} d\alpha_j = 0 \tag{6.16}$$

Solving the above gives

$$\frac{d\alpha_j}{d\mu_i} = -\frac{\frac{\partial E[p^*]}{\partial \mu_i}}{\frac{\partial E[p^*]}{\partial \alpha_j}} \tag{6.17}$$

or

$$\frac{d\alpha_j}{d\mu_i} = -\frac{1}{\mathbf{1}^T(\mathbf{I}-\rho\mathbf{W})^{-1}\boldsymbol{e}_1} \tag{6.18}$$

This equation allows to compute the amount of renewables $d\alpha_j$ needed to restore the pre-change price level.

Notice that the absence of spatial correlation in zonal demand ($\rho = 0$) implies $\frac{d\alpha_j}{d\mu_i} = 1$, namely, perfect proportionality between the size of the change and the size of the adjustment in parameter α_j required to sterilize the price increase. $\rho > 0$ instead would make the adaptation effort more expensive: for instance, more renewable energy capacity would have to be installed or higher incentives for the supply of renewables would have to be given.

The value $d\alpha_j$ determined through this analysis would be the outcome of policies that foster investments in new renewable energy capacity or that slows down their market penetration. $d\alpha_j$ reflects the notion of recoverability, since a higher required change in renewable energy capacity takes time to be implemented. Indeed, this depends on the length of policy-making and investment decision-making processes, as well as on time for new capacity building. The sum of the two marks the duration of the recovery period.

The outlined theoretical framework sheds light on a novel interpretation of the existing estimates of spatial econometric models of wholesale power demand. Previous estimates of the spatial correlation coefficient can prove useful in order to assess the potential resilience of the power systems analysed in the literature, which we will review in the next subsection.

6.3.2 *Literature Review*

A few empirical studies on electricity demand have recognized the role of spatial interactions.[9] In this strand of literature, the impact of climate shocks on power demand is most often implicit, as weather is treated as unobservable, collapsed in the error term along with other disturbances. Though the robust evidence on a positive spatial autocorrelation in power demand indicates, in light of the above theoretical analysis, that shocks are amplified along their spatial propagation process and may make resilience more costly and the recovery longer.

The next table (Table 6.1) presents a summary of the existing estimates of spatial autocorrelation in power demand. For each cited paper, the table reports baseline information on the dataset and on model specifications. In particular, the table only refers to the "best" model, according to specification tests, by the authors.

As it can be spotted from the table, through different modelling strategies the cited papers all come to the conclusion that energy consumption at the provincial or regional level is spatially autocorrelated (ρ in Eq. 6.7). Point estimates of the coefficients range from about 16% (Turkish provinces in Akarsu 2017) to about 77% (Chinese provinces in Yu et al. 2012), signalling a rather wide heterogeneity in the underlying energy systems and regional economies, but perhaps also in data collection methods, in data quality, and in the adopted econometric specifications.[10] Gathering all the data used by the authors and running the same model specification and tests would be an interesting comparative experiment, provided that all explanatory variables for the same time window and spatial resolution be available. Such a unified analysis or other meta-analytical techniques would perhaps be needed before any comparative assessment. In addition, it must be noted that, while some of the reviewed papers include weather variables as regressors, none of them concentrates on a specific climate-related shock. In other words, the error terms in the estimated

[9]This section partly builds on De Siano and Sapio (2020).

[10]The set of explanatory variables included in the spatial econometric models differs across the reviewed papers. The focus of this section, though, is on spatial autocorrelation, and hence we do not describe the coefficient estimates associated to control variables, which are mostly in line with theoretical expectations.

models conflate shocks generated by different sources. Furthermore, the choice of a specific spatial resolution for the analysis is most often guided or constrained by data availability, and may not coincide with the spatial resolution of policy responsibilities. A high spatial autocorrelation at the provincial level may have relevant policy implications, but the appropriate policy tools may be under the responsibility of higher level administrative bodies (say, regional or national). With these caveats in mind, the purpose of illustrating the analytical framework presented in Sect. 6.3 is served by noting that the stronger spatial autocorrelation in Chinese and Spanish power consumption implies larger resource requirements than in Turkey to restore the pre-shock energy costs in case the system is hit by a shock of a given size (be it a climate related or other shock). Further investigations should be performed in order to identify the institutional or technological reasons behind such differences.

The reviewed works also offer useful illustrations of the methodological steps described in Chap. 3 of this book, in particular, as regards Exploratory Spatial Data Analysis (Sect. 3.3) and Specification Tests (3.4.3).

The first methodological step in all the papers considered here involved a test-based motivation for using a spatial econometric approach. Moran's *I* statistic has been generally used to provide evidence of spatial interdependency. The results of the ESDA in Yaylaci et al. (2011) reveal the presence of a positive spatial autocorrelation based on global and local spatial indicators (Moran's I, Gi, and Gi* statistics). Gi* statistics above 1.96 are observed in Istanbul and in four other cities. In Akarsu (2017), the test statistic is separately computed for each spatial aggregation level (NUTS-2, NUTS-3), and in specific years, such as the first and last sample year. The value of the test statistic varies across years and aggregation levels (between 0.2521 and 0.4547), but is in all cases statistically significant at conventional levels. Moran's *I* in Yu et al. (2012) equals 0.2360. Blazquez Gomez et al. (2013) additionally perform the conditional Lagrange multiplier tests described in Baltagi et al. (2003) and reject the null hypothesis of no spatial autocorrelation.

Specification tests allow to select the best fitting spatial econometric specification among the available ones. In the spatial econometric literature, theory is seldom called upon for guidance on specification issues, and hence the procedures being followed are purely empirical. Blazquez Gomez et al. (2013), for instance, conclude that a SARAR model (spatial autoregressive with autoregressive disturbances) is justified upon comparing the results of a conditional Lagrange multiplier test for the null of no spatial autocorrelation (Baltagi et al. 2003) with a conditional Lagrange multiplier test due to Baltagi and Long (2008) for the null of no spatial lag dependence. Yu et al. (2012) find that the SAR model specification outperforms the SEM based on R-squared values. Their approach relies on the Gibbs sampling method based on the Markov Chain Monte Carlo (MCMC) model to construct a Bayesian spatial econometric model. This allows to address potential heteroscedasticity problems due to geographic, economic, societal, scientific, technological, population and cultural differences that characterize Chinese provinces. In papers applying SAR or SAR-ARMA models, one of the key specification steps involves lag selection, which is performed through the likelihood test procedures ordinarily available for data organized as time series.

Table 6.1 Summary of spatial econometric estimates of spatial autocorrelation in power demand. Legend: ρ denotes the point estimate of the spatial autoregressive coefficient; PD: panel data; DSD: dynamic spatial Durbin

Authors (year)	Country	Period	Unit	*W*	"Best" model	ρ
Ohtsuka et al. (2010)	Japan	1992–2003	Regions	Binary contiguity	SAR-ARMA	0.3827
Yaylaci et al. (2011)	Turkey	2006	Provinces	Binary contiguity	(only ESDA)	
Yu et al. (2012)	China	2007–2009	Provinces	Binary contiguity	SAR	0.7699
Blazquez Gomez et al. (2013)	Spain	2001–2008	Provinces	Inverse distance	SARAR-PD	0.6770
Ohtsuka and Kakamu (2013)	Japan	1992–2003	Regions	Binary contiguity	SAR-ARMA	
Akarsu (2017)	Turkey	1990–2001	Provinces (NUTS-3)	Binary contiguity	DSD-PD	0.1583
		1990–2001	Regions (NUTS-2)			0.2400
		2004–2011	Regions (NUTS-2)			0.2973

Finally, two useful remarks are inspired by the analysis in Ohtsuka and Kakamu (2013). In their paper, regions are considered contiguous if they are connected by overhead transmission lines or undersea cables, even if they do not actually share a border (as with islands). Moreover, and rather uncommon in the spatial econometric literature, selection of the best performing model is ultimately based on forecasting performance. Interestingly, the authors consider VAR models as "competitors" of the SAR-ARMA model in forecasting terms (Table 6.1).[11]

[11]De Siano and Sapio (2020) comment on the advantages of spatial econometrics versus vector autoregressions in testing hypotheses about electricity markets, such as market integration. The extent of spatial effects cannot be easily inferred from the existing multivariate time-series estimates on electricity prices, because all locations in a VAR model are treated equally, regardless of physical or socio-economic distance. Moreover, time is unidirectional and linear, whereas space has no obvious beginning and direction, and any location can have more than one neighbour.

6.4 The Adoption of Renewable Energy Technologies

A long-term perspective on the resilience of power systems to climate shocks involves ecological innovation (Kemp 2010) and the transition to clean energy sources. Recent decades have been characterized by the global deployment of renewable energy, mainly driven by concerns over climate change, threats to the security of energy imports and the price risk induced by market deregulation. The role played by households in the energy transition may be crucial, as the move from "dirty" to "clean" energy sources is also a shift away from the centralized energy generation paradigm to a decentralized paradigm, wherein household has the capability to generate the energy they need (Künneke 2008). With adequate legislative and financial support, households may change their role in the industry by becoming self-producers (*prosumers*) and have a hands-on approach to the consequences of climate shocks. Prosumers can promote the resilience of the power industry to climate-induced damages to centralized power generation facilities. Indeed, a temporary curtailment of supply from large-scale power plants can be offset through an increase in self-generation by prosumers or by an increasing injection of power by prosumers into the grid.

Among the renewable energy technologies, photovoltaic (PV) panels are the most suitable for adoption by small-scale energy-using facilities. Quite a rich empirical literature on the uptake of PV panels has emerged (Bollinger and Gillingham 2012), highlighting the importance of market-pull policies and feed-in-tariff schemes to overcome barriers that slow down their diffusion, such as high overhead costs, low predictability or supply curtailments.[12] Within the literature on PV uptake, a stream of papers have therefore studied the spatial correlation in adoption decisions by households. The next section outlines mechanisms that would imply spatial effects and theoretical insights that predict the existence of spatial correlation in new technology adoption.

6.4.1 *Theoretical Insights*

The literature on technology diffusion (e.g. Rogers 1983) has long recognized the role of spatial proximity in promoting progress along an S-shaped diffusion curve, whereas recent advances in cognitive economics provide further motivations for a spatial framework of analysis.

In illustrating the theoretical insights behind spatial econometric models of PV adoption, it is useful to distinguish between peer effects and spatial spillovers. Peer effects, in turn, can refer to active peer effects when individuals persuade someone

[12] See also the agent-based simulation model by Palmer et al. (2015), who have studied the diffusion of PV generation systems under different support schemes, and calibrated the model on Italian data.

else to use the same electrical power facility and passive peer effects when the decision to adopt a renewable energy technology is made after observing what the neighbours do. Spatial spillovers arise when peer effects spread across localities.

While active peer effects may have to do with profit seeking, there are several ways of rationalizing passive peer effects:

- Information can be classified as codified and tacit (Polanyi 1996). Tacit information, which is often critical for the value of an innovation, can only be transmitted through individual contacts, and hence it requires spatial proximity.
- The normative expectations that lead individuals to feel responsible for the environment (awareness of consequences according to Schwartz 1977) may stem from interactions with neighbours, such as family members, friends or the active engagement in social networks (Gadenne et al. 2011; Martinsson et al. 2011).
- Impure altruism or warm glow effects (Andreoni 1990) arise when individuals enjoy being recognized as altruists, e.g. with respect to the environment. Such effects may push some individuals to make their "green" technology adoption decision visible to their neighbours, thereby influencing their choice.
- Saliency effects may partly explain why decisions made by close neighbours may have a stronger weight on an individual, than decisions made by farther neighbours.

The above mechanisms may be strengthened or mitigated by geography and urbanization. Geography plays a key role in solar exposition and the corresponding average annual solar radiation detected in a place, which both affect the efficiency and profitability of PV installations. Denser urban areas facilitate information transmission between PV adopters and their peers (direct network effects) or ensuring a greater presence of skilled workers employed in the PV sector (indirect network effects).

An early modelling tradition (Mansfield 1961) was based on an analogy between technology diffusion and the diffusion of an epidemic. Such analogy has inspired a number of recent investigations in spatial econometrics to use an epidemic diffusion model.

In general, an epidemic diffusion model can be described by a population N and the number of users at time t, y_t (see also Geroski 2000). The number of new adopters is measured by Δy_t and can be modelled as a non-linear function of past values of y:

$$\Delta y_t = [\alpha + \beta y_{t-1}][N - y_{t-1}] \tag{6.19}$$

where $\alpha \in [0, 1]$ is the coefficient of external influence, i.e. the number of non-users that adopt the new technology regardless of information from the existing users. The coefficient β, instead, is the coefficient of internal influence and measures the extent to which existing users are "imitated".

The basic epidemic model is extended by Rode and Weber (2016) on the insight that spatial proximity matters because of the tacit component in information and knowledge. This is done by assuming that the number of new adopters in location i reads

$$\Delta y_{it} = [\alpha + \beta_0 y_{i,t-1} + \beta_1 y_{i,1,t-1} + \cdots + \beta_Q y_{i,Q,t-1}][N - y_{t-1}] \quad (6.20)$$

where $y_{i,x,t-1}$ is the number of existing adopters within a certain distance x from the location of firm i and β_x is the associated coefficient. The above is essentially a dynamic spatial econometric model, wherein the β_x coefficients collapse information from the spatial autocorrelation coefficient and from the entries of the spatial-weights matrix.

6.4.2 *Literature Review*

The theoretical insights on spatial patterns in technology adoption can be useful in interpreting the spatial econometric evidence on the uptake of PV panels in the residential energy consumption sector, reviewed below.

Dharshing (2017) finds strong peer effects and neighbouring spillovers in household solar panel installation. The author considers a panel dataset that includes almost one million PV systems in 402 German counties, observed in the 2000–2013 time window. County-level spatial proximity is taken into account by applying a row-standardized rook binary contiguity matrix. The presence of spatial dependence between neighbouring counties is confirmed by the results of Moran's-I global indicator. Estimates of SAR and SEM models then reveal the presence of strong spatial dependence that extends beyond the effects of socio-economic and demographic factors. Similar results for Germany are found by Schaffer and Brun (2015) who show, by means of a SAR panel model, that patterns of PV diffusion vary between counties due to neighbourhood effects, resulting from an uneven diffusion of skills and intermediary agents, controlling for spatial heterogeneity in solar radiation expositions.

Works focusing on UK data have incorporated information on policies and assessed their effects. Snape (2016) accounts for spatial effects in a descriptive analysis by applying a mixed strategy based on quantitative data on PV adoption and qualitative information on energy policies applied from 2010 to 2015. The use of a global spatial indicator (Moran's-I index) reveals the presence of positive spatial autocorrelation at the post-code district level. Instead, local indicators (LISAs) describing the geographical positioning and strength of clusters reveal clusters of high intensity capacity in southwestern England and highlight the emergence of a positive spatial autocorrelation in eastern England. The analysis confirms the presence of low-low clusters around large cities like London, Birmingham, Manchester and Liverpool. Richter (2013) finds that passive peer effects strongly influence further installation decisions at a post-code area level of geographical disaggregation.

The focus in Balta-Ozkan et al. (2015b) is on the effects of feed-in tariffs on household PV adoption in British regions. After positive spatial dependence is detected through Moran's-I test, the authors build a spatial econometric model using an inverse distance weight matrix, and including control variables such as per capita income, population, house density, environmental conditions, electricity demand and solar irradiation levels. A general-to-specific approach (Elhorst 2010a, b) based on Lagrange multiplier tests leads to select the SDM as the most appropriate model to describe PV deployment. According to the results, the dependent variable (PV uptake) as well as the explanatory variable is responsible for spatial spillovers. To account for potential endogeneity biases, a generalized spatial two-stage least squares procedure (GS-2SLS) is applied while using spatially lagged explanatory variables as instruments. The SDM and GS-2SLS results turn out to be similar.

In the perspective of learning the practice of spatial econometrics, the paper by Dharshing (2017) is well suited for an illustration. Let us then proceed and describe the methodological steps followed by the author, echoing the contents of Chaps. 3 and 4 in this book. In doing so, we are more interested in guiding the reader through the steps than in commenting specific results.

At the outset of the paper, the author motivates the analysis from both academic and policy-making perspectives against the background of an increasing diffusion of residential PV technologies, and positions his work within an emerging stream of literature attempting to understand the factors behind the observed spatial patterns. The value added of the proposed analysis lies precisely in the adoption of spatial modelling techniques that are expected to overcome the limitations of previous econometric approaches (such as fixed effects panel models, binary panel logit models and non-spatial epidemic diffusion models; see Table 1 in Dharshing 2017).

Next, the dataset is described. The most crucial decisions at this stage concern the spatial resolution of the analysis, the measurement of the dependent variable and the selection of the explanatory variables (besides the spatial interaction terms). The paper focuses on county-level data (approximately corresponding to NUTS-3). This choice seems to be a good solution in terms of information content, since data at higher aggregation levels would be less informative, while county-level data satisfy a data availability constraint (even more disaggregate data would be difficult and costly to obtain). Yet, this is not without measurement challenges, since administrative district boundaries changed along the sample period, forcing the author to harmonize the data. The dependent variable is chosen in line with the goal of the analysis and in a way that singles out the effect of new installations. Therefore, the author focuses on the number of residential PV installations, restricted to small-scale rooftop PV systems of between 1 and 10 kWp, divided by the number of owner-occupied buildings, after subtracting the number of all buildings with pre-existing PV installations. The data is retrieved from the four major grid operators in Germany. Data cleaning leaves 589,202 PV systems available for the analysis. Explanatory variables are categorized as (i) adopter characteristics, (ii) economic profitability of PV systems and (iii) settlement structure. Interestingly, the inclusion of proxies for the economic profitability of PV systems, such as return on investment (ROI), is motivated through a residential investment model. Still, the theoretical model does not imply any restriction on the functional form of the relationship between PV diffusion and its determinants.

Alternative models previously used in the literature, focusing on payback period, net present value or the internal rate of return, are however cited.

Further decisions regard model specification. This involves two issues: the choice of a spatial-weights matrix and then the choice among the various spatial econometric models available in the literature. The author follows common practice in the literature by choosing a rook-binary contiguity matrix, row-standardized to unity. There is no specific theoretical reason behind this choice. The author follows a pragmatic approach, according to which the results based on the spatial weight matrix of choice are subject to robustness checks by using an alternative—and a priori not less sensible—spatial weight matrix. Once a decision on the matrix W is made, the candidate models are a SAR panel model and a SEM panel model. Again, the author prefers to let the data speak rather than imposing a theory-driven model specification. Testing procedures such as Moran's I and the Baltagi et al. (2003) conditional LM test show that spatial econometric models are justified, confirming the insight behind the paper. A Hausman test points to a fixed effects specification. The model is estimated via a ML approach, following the industry standard, although the possibility of running GMM is mentioned.

Finally, after illustrating the results, the author spends some time in discussing some methodological issues that allow to grasp limitations and potential improvements of the analysis. First, a robustness exercise using an alternative spatial-weights matrix (based on a Delaunay triangulation approach) is performed, showing that results do not change significantly. Second, dynamic spatial models are suggested as providing a solution to the reflection (or simultaneity) problem that may hamper the analysis. Third and last, some thoughts are devoted on the issue of spatial granularity, much in the spirit of Sect. 6.2 in this chapter.

6.5 Conclusion

In a time when climate shocks and the related damages are expected to grow in frequency and intensity, spatial econometrics has not yet become the mainstream approach in power system economics despite the threatening consequences of shock propagation through space. The evidence reviewed in this chapter and the theoretical insights on spatial effects will hopefully ameliorate our understanding of the power industry resilience to climate shocks, whether through endogenous forces or with the aid of appropriate policy measures.

The chapter has also explored some practical issues that applied researchers can face in their econometric design, such as measurement issues, the spatial granularity of the data and the criteria for choosing the spatially weighted matrices that are relevant for power demand and the diffusion of renewable energy technologies.

So far, spatial econometric research on power demand has faced some methodological limitations. The most promising avenues for improvement are the following. First, the simultaneous inclusion of serial and spatial correlation terms in the same model would allow to take care of persistence in power demand and at the same time to create a link with the energy finance community and its expertise in risk modelling and assessment. Persistence in power demand and the ensuing serial correlation is particularly sizeable in higher frequency datasets (e.g. daily data), as patterns of power consumption are habitual, and retail electricity prices display very little variability over such a short time horizon.

In addition, specifying a serial correlation structure alongside the spatial correlation terms could allow to identify temporal chains of causation, a highly critical issue as some locations may be more exposed to climate shocks, or more central in the power network. Leader-follower dynamics are epitomized by events, such as the temporary suspension of 18 nuclear power plants in France for precautionary reasons, whose effects propagated throughout Europe (Rinne 2019; Sapio 2016).

Finally, advances in the econometric approach to the resilience of power systems could descend from investigating spatial patterns in power supply. An extension of the theoretical analysis outlined in Sect. 6.3.1 should incorporate spatial correlations also on the supply side of the market and assume that also the power generation facilities are exposed to climate shocks, both in terms of resource availability (e.g. wind speeds increasing due to climate change) and in terms of physical damages to generation units. To our knowledge, only few applications of spatial econometrics to renewable energy have been published, notably by Croonenbroeck and Ambach (2015) on wind power and by Bowen and Lacombe (2017) who have taken spatial autocorrelation into account in assessing the impact of renewable portfolio standards on renewable energy generation.

References

Akarsu, G. (2017). Analysis of regional electricity demand for Turkey. *Regional Studies, Regional Science*, *4*(1), 32–41.

Andreoni, J. (1990). Impure altruism and donations to public goods: A theory of warm-glow giving. *The economic journal*, *100*(401), 464–477.

Auffhammer, M., & Mansur, E. T. (2014). Measuring climatic impacts on energy consumption: A review of the empirical literature. *Energy Economics*, *46*, 522–530.

Balta-Ozkan, N., Watson, T., & Mocca, E. (2015a). Spatially uneven development and low carbon transitions: Insights from urban and regional planning. *Energy Policy*, *85*, 500–510.

Balta-Ozkan, N., Yildirim, J., & Connor, P. M. (2015b). Regional distribution of photovoltaic deployment in the UK and its determinants: A spatial econometric approach. *Energy Economics*, *51*, 417–429.

Baltagi, B. H., Long, L. (2008). *Testing for random effects and spatial lag dependence in panel data models*. Center for Policy Research. Working Paper No 102.

Baltagi, B. H., Song, S. H., & Koh, W. (2003). Testing panel data regression models with spatial error correlation. *Journal of Econometrics*, *117*(1), 123–150.

Blazquez Gomez, L. M., Filippini, M., & Heimsch, F. (2013). Regional impact of changes in disposable income on Spanish electricity demand: A spatial econometric analysis. *Energy Economics*, *40*(Suppl. 1), S58–S66.

Boffa, F., & Sapio, A. (2015). Introduction to the special issue "The regional integration of energy markets". *Energy Policy*, *85*, 421–425.

Bohn, R. E., Caramanis, M. C., & Schweppe, F. C. (1984). Optimal pricing in electrical networks over space and time. *Rand Journal of Economics*, *15*(3), 360–376.

Bollinger, B., & Gillingham, K. (2012). Peer effects in the diffusion of solar photovoltaic panels. *Marketing Science*, *31*(6), 900–912.

Bonjean Stanton, M. C. B., Dessai, S., & Paavola, J. (2016). A systematic review of the impacts of climate variability and change on electricity systems in Europe. *Energy*, *109*, 1148–1159.

Bowen, E., & Lacombe, D. J. (2017). Spatial dependence in State renewable policy: Effects of renewable portfolio standards on renewable generation within NERC regions. *Energy Journal*, *38*(3), 177–193.

Bresnahan, T. F., & Trajtenberg, M. (1995). General purpose technologies 'Engines of growth'? *Journal of Econometrics*, *65*(1), 83–108.

Carmona, R., & Coulon, M. (2014). A survey of commodity markets and structural models for electricity prices. In F. Benth, V. Kholodnyi, & P. Laurence (Eds.), *Quantitative energy finance* (pp. 41–48). New York: Springer.

Chandramowli, S. N., & Felder, F. A. (2014). Impact of climate change on electricity systems and markets-a review of models and forecasts. *Sustainable Energy Technologies and Assessments*, *5*, 62–74.

Croonenbroeck, C., & Ambach, D. (2015). Censored spatial wind power prediction with random effects. *Renewable and Sustainable Energy Review*, *51*, 613–622.

De Siano, R., & Sapio, A. (2020). Spatial Econometrics in Electricity Markets Research. In D.-T. Nguyen & S. Goutte (Eds.), *Handbook of energy finance: theories*. Practices and simulations: World Scientific Publishing.

Dharshing, S. (2017). Household dynamics of technology adoption: A spatial econometric analysis of residential solar photovoltaic (PV) systems in Germany. *Energy Research & Social Science*, *23*, 113–124.

Elhorst, J. P. (2010a). Spatial panel data models. In M. M. Fischer & A. Getis (Eds.), *Handbook of applied spatial analysis*. Berlin: Springer.

Elhorst, J. P. (2010b). Applied spatial econometrics: Raising the bar. *Spatial Economic Analysis*, *5*, 9–28.

Gadenne, D., Sharma, B., Kerr, D., & Smith, T. (2011). The influence of consumers' environmental beliefs and attitudes on energy saving behaviours. *Energy policy*, *39*(12), 7684–7694.

Geroski, P. A. (2000). Models of technology diffusion. *Research policy*, *29*(4–5), 603–625.

Granovetter, M., & McGuire, P. (1998). The making of an industry: Electricity in the United States. *The Sociological Review*, *46*(S1), 147–173.

Green, R. (2007). Nodal pricing of electricity: How much does it cost to get it wrong? *Journal of Regulatory Economics*, *31*(2), 125–149.

Helm, D., & Hepburn, C. (2019). The age of electricity. *Oxford Review of Economic Policy*, *35*(2), 183–196.

Kemp, R. (2010). Eco-Innovation: definition, measurement and open research issues. *Economia politica*, *27*(3), 397–420.

Künneke, R. W. (2008). Institutional reform and technological practice: The case of electricity. *Industrial and Corporate Change*, *17*(2), 233–265.

Mansfield, E. (1961). Technical change and the rate of imitation. Econometrica: *Journal of the Econometric Society*, 741–766.

Martinsson, J., Lundqvist, L. J., & Sundström, A. (2011). Energy saving in Swedish households. The (relative) importance of environmental attitudes. *Energy policy*, *39*(9), 5182–5191.

Nijkamp, P. (1983). Regional dimensions of energy scarcity. *Environment and Planning C: Government & Policy*, *1*(2), 179–192.

Ohtsuka, Y., & Kakamu, K. (2013). Space-time model versus VAR model: Forecasting electricity demand in Japan. *Journal of Forecasting, 32*, 75–85.

Ohtsuka, Y., Oga, T., & Kakamu, K. (2010). Forecasting electricity demand in Japan: A Bayesian spatial autoregressive ARMA approach. *Computational Statistics & Data Analysis, 54*, 2721–2735.

Palmer, J., Sorda, G., & Madlener, R. (2015). Modeling the diffusion of residential photovoltaic systems in Italy: An agent-based simulation. *Technological Forecasting and Social Change, 99*, 106–131.

Panteli, M., & Mancarella, P. (2015). Influence of extreme weather and climate change on the resilience of power systems: Impacts and possible mitigation strategies. *Electric Power Systems Research, 127*, 259–270.

Polanyi, M. (1966). *The tacit dimension*. University of Chicago press.

Richter, L. L. (2013). *Social effects in the diffusion of solar photovoltaic technology in the UK*. Cambridge, UK: University of Cambridge.

Rinne, S. (2019). Radioinactive: Do nuclear power plant outages in France affect the German electricity prices? *Energy Economics*, 104593.

Rode, J., & Weber, A. (2016). Does localized imitation drive technology adoption? A case study on rooftop photovoltaic systems in Germany. *Journal of Environmental Economics and Management, 78*, 38–48.

Rogers, E. M. (1983). Diffusion of Innovations, A Division of Macmillan Publishing Co. *Inc. Third Edition, The Free Pres, New York*.

Sapio, A. (2016). Reattori spenti in Francia: quanto ci costano? https://rienergia.staffettaonline.com/articolo/32727/Reattori+spenti+in+Francia:+quanto+ci+costano/Sapio.

Schaffer, A. J., & Brun, S. (2015). Beyond the sun? Socioeconomic drivers of the adoption of small-scale photovoltaic installations in Germany. *Energy Research & Social Science, 10*, 220–227.

Schaeffer, R., Szklo, A. S., de Lucena, A. F. P., Borba, B. S. M. C., Nogueira, L. P. P., Fleming, F. P., et al. (2012). Energy sector vulnerability to climate change: A review. *Energy, 38*(1), 1–12.

Schwartz, S. H. (1977). Normative influences on altruism. *Advances in experimental social psychology, 10*(1), 221–279.

Snape, J. R. (2016). Spatial and temporal characteristics of PV adoption in the UK and their implications for the smart grid. *Energies, 9*(3), 210.

Walton, S., & Tabors, R. D. (1996). Zonal transmission pricing: methodology and preliminary results from the WSCC. *The Electricity Journal, 9*(9), 34–41.

Yasner, R. A. (2012). Maximizing Renewable Electricity in Israel: Energy Security, Environmental Impact, and Economic Development. Doctoral dissertation, Carnegie Mellon University.

Yaylaci, E. D., Ismaila, A. B., Uskay, O., & Düzgün, S. (2011). Spatial analyses of electricity supply and consumption in Turkey for effective energy management and policy-making. In M. Schmidt, V. Onyango, & D. Palekhov (Eds.), *Implementing environmental and resource management* (pp. 153–168). Berlin-Heidelberg: Springer.

Yu, Y., Li, X., Qian, Z., Yang, M., Li, Z., Guo, P., Liu, L., & Zhu, X. (2012) A study on China's energy consumption by Bayesian spatial econometric model. *Energy Procedia, 16*, 1332–1340.

Chapter 7
Conclusion and Open Issues

Abstract This concluding chapter critically illustrates the evolution of the empirical analysis of regional resilience with respect to climatic shocks. The importance of having considered the presence of spatial interdependencies in analysing the response of regions to different types of external shocks is underlined. However, the authors also briefly define the new challenges for research in this field. First of all the construction of endogenous distance matrices, capable of re-modeling the interdependencies according to the changing effects the shocks exert on the observed geographical units. Another issue is represented by the need for developing new methods that may allow to explore the strength of spatial spillovers and peer effects under future different scenarios. Despite these caveats, finding appropriate ways of exploiting climate-economy projections can enhance the policy relevance of climate econometrics in general—and spatial climate econometrics in particular.

Keywords Resilience · Climate shocks · Spatial econometrics · Social interactions · New challenges · Research challenges · Endogeneity · Scenarios

The main ambition of this book was to collect, illustrate and critically evaluate up-to-date methods and results on the economic resilience of regions with respect to climate-related shocks. The starting point in this process was the acknowledgment that most econometric studies on resilience have overlooked the significance—in policy terms—of estimating spatial spillovers and peer effects. At the same time, the very process of reviewing the literature has allowed us to identify stimulating new avenues for research on spatial climate change econometrics.

The early empirical analyses on regional resilience focused mainly on the idiosyncratic—and merely local—characteristics that would make regions more resilient. The main aim was to help policy-makers in building measures and strategies able to reduce spatial vulnerability of the system to shocks and/or to enhance their ability in facing and recovering from a crisis (Bristow and Healy 2018; Crespo et al. 2014; Wink 2014). Empirical analyses so far relied almost exclusively on a time-series approach to estimate and forecast regional resilience, treating

R. De Siano et al., *Regional Resilience to Climate and Environmental Shocks*, SpringerBriefs in Regional Science, https://doi.org/10.1007/978-3-030-54588-8_7

geographic units as "isolated islands". This constitutes a weakness of these investigations as they completely overlook the possibility of spatial interactions affecting regional capacities to recover from shocks and the redesign of their policies. Despite the linkages between nearby territories have long been a key issue for geographers (Tobler 1970), previous studies do not verify whether low (high)-resilient regions are arbitrarily distributed or tend to cluster together due to the emergence of spatial interdependencies. Not enough relevance is given to the question: to what extent a region's responsiveness to economic shocks depends on the neighbouring ones—or if it can condition them. Yet, the worldwide changes of the last decades show that countries, and even more regions, are strongly connected to each other. Shocks occurring in a given territory and policies implemented as a response eventually spread their effects over neighbouring areas. Positive and negative externalities affecting neighbouring countries/regions may contribute to change resilience highlighting the inadequacy of traditional linear modelling as already showed in the econometric literature with regard to other economic issues.

The various technical arguments in favour of spatial methods, presented by LeSage and Pace (2009), highlight shortcomings in non-spatial estimators that cast doubts on the robustness of the empirical results and may undermine the credibility of any intervention aimed at strengthening regional economic resilience. Differences in regional resilience may be surely explained by the specific abilities of local policy-makers to build effective policy responses, but, at the same time, the local performance of policies can be enhanced or hampered by spatial interactions with neighbouring geographical areas, beyond the intrinsic merits or faults of the local politicians.

The availability of localized data, combined with the spatial statistics procedures pre-programmed into multiple statistical software tools, raises the question of how this proximity could be modelled into territorial economic studies.[1] Thus, regarding regional resilience, the availability of new econometric tools for modelling spatial dependence provides researchers with suitable instruments to correctly attribute successes and failures among local policy-makers, thereby helping to shape resilient behaviours and to develop more effective policies.

Relatedly, an inherent difficulty in the analysis rooted in geographical information systems is that geographic effects are often mediated by social interactions. The spatial diffusion of "best practices" in policy-making is an instance of how social interactions matter, as well as the geographical spread of information on the virtues of certain policy solutions. Social interactions in such examples yield effects on more levels: one could envisage that a region's resilience to shocks is correlated to its neighbours' resilience because policy-makers in one region "imitate" policy decisions made by apparently successful neighbouring regions. Unfortunately, the study of such interplay between spatial and social/economic relations is still an under-researched topic. There are multiple reasons for that. One is the discipline divide that often exists between the researchers dealing with geographical data and those investigating social networks. This makes it difficult to create a shared methodolog-

[1] LeSage and Pace (2009), Elhorst (2003, 2010) for MATLAB; GeoDa, by Anselin (2003); for Stata, Belotti et al. (2017) among others; for R, Bivand et al. (2018), Millo and Piras (2017).

ical toolbox able to analyse socio-spatial dynamics. Another reason is the practical difficulty to work with software allowing to manage both spatial and social network data. In Chap. 4, we aim at providing one of the first contributions to fill the gap between the two disciplines, by showing how it is possible to model spatial and social/economic relations. We hope that the techniques discussed in the chapter will be soon extended for future studies.

However, spatial econometric procedures present some methodological limitations and pose challenges for the theoretical and empirical literature. Overcoming limitations and responding to challenges become relevant also when modelling space in resilience analyses. A first issue that concerns characterizing proximity between local units is usually based on the geographical distance or different types of aggregations according to social, economic and cultural features. In most cases, the empirical analysis tends to be conditioned by an a priori definition of the neighbourhood, whereas one should account for any change that could impact on neighbourhood relations. The initial choice of the spatial weight matrix and the possibility of adjusting it are crucial points in the spatial analysis because the interpretation of the estimators as well as the nature and size of the spatial dependence strongly depend on them. Some progresses have been made in this regard. Indeed, while one of the most important requirements has so far been the use of an exogenous spatial weight matrix, recently Kelejian and Piras (2014) introduced a more realistic approach allowing for the use of matrices with weights that are obtained endogenously.

This issue emerges rather strongly when dealing with climate-induced shocks, since climate change may alter the strength of spatial linkages. This is clearly problematic for research works that measure spatial linkages through output measures: if their results depend critically on the entries of the spatial-weights matrix, their "out-of-sample" validity vanishes once climate impacts on regional outputs are considered, at least on a rather long time horizon. Distance-based spatial linkages, though, are no less exposed to the modelling risks related to long-term endogeneity. For instance, climate shocks that permanently damage transport infrastructures may make two cities farther than they currently are.

The above considerations point to the need for developing new methods that may allow to explore the strength of spatial spillovers and peer effects under future scenarios. Scenarios may consist of alternative assumptions on the future values assumed by the spatial-weights matrix, under different climatic projections. Incidentally, this would also free climate econometrics from its narrowly retrospective outlook. In this respect, Chap. 5 has illustrated the merits and demerits of the projections based on standard IAMs and the value added from agent-based IAMs. Climate-economy projections can also help identifying the most salient weather extremes and the most exposed locations, with remarkable advantages in the process of sample selection and in the ability to pick convincing case studies. Perhaps, the value added of such dialogue between simulation and econometric models can be even magnified in the assessment of climate impacts on the power industry, as power market simulations are typically based on models that incorporate faithful representations of the transmission network. Different nodes in the power grid are exposed to different local climate conditions and may face different risks or opportunities (see Chap. 6). Caution is required, as the high (low) quality of projections would spill over to the high (low) quality of policy prescriptions that can be extrapolated from spatial climate econometric estimates. Despite these caveats, finding appropriate ways of exploiting climate-economy projections can enhance the policy relevance of climate econometrics in general—and spatial climate econometrics in particular.

References

Anselin, L. (2003). GeoDa 0.9 user's guide. Spatial Analysis Laboratory (SAL). Department of Agricultural and Consumer Economics, University of Illinois, Urbana-Champaign, IL.

Anselin, L., Bera, A. K., Florax, R., & Yoon, M. J. (1996). Simple diagnostic tests for spatial dependence. *Regional Science and Urban Economics*, *26*(1), 77–104.

Belotti, F., Hughes, G., & Mortari, A. P. (2017). Spatial panel-data models using Stata. *The Stata Journal*, *17*(1), 139–180.

Bivand, R., Altman, M., Anselin, L., Assunçao, R., Berke, O., Bernat, A., Blanchet, G., Blankmeyer, E., Carvalho, M., Christensen, B., Chun, Y., Dormann, C., Dray, S., Gomez-Rubio, V., Gubri, M., Halbersma, R., Krainski, E., Legendre, P., Lewin-Koh, N., Li, H., Ma, J., Mallik, A., Millo, G., Mueller, W., Ono, H., Peres-Neto, P., Piras, G., Reder, M., Tiefelsdorf, M., Westerholt, R., & Yu, D. (2018). SPDEP: Spatial dependence: Weighting schemes, statistics and models. https://cran.r-project.org/web/packages/spdep/.

Bristow, G., & Healy, A. (Eds.). (2018). *Economic crisis and the resilience of regions: A European study*. Cheltenham: Edward Elgar.

Crespo, J., Suire, R., & Vicente, J. (2014). Lock-in or lock-out? How structural properties of knowledge networks affect regional resilience. *Journal of Economic Geography*, *14*(1), 199–219.

Elhorst, J. P. (2003). Specification and estimation of spatial panel data models. *International Regional Science Review*, *26*, 244–268.

Elhorst, J. P. (2010). Spatial panel data models. In M. M. Fischer & A. Getis (Eds.), *Handbook of applied spatial analysis* (pp. 377–407). Berlin: Springer.

Kelejian, H. H., & Piras, G. (2014). Estimation of spatial models with endogenous weighting matrices, and an application to a demand model for cigarettes. *Regional Science and Urban Economics*, *46*, 140–149.

Le Gallo, J. (2002). Econometrie spatiale: l'autocorrelation spatiale dans les modeles de regression lineaire. *Economie & Prevision*, *4*, 139–157.

LeSage, J., & Pace, R. K. (2009). *Introduction to spatial econometrics*. Chapman and Hall/CRC.

Millo, G., & Piras, G. (2017). SPLM: Econometric models for spatial panel data. https://cran.r-project.org/web/packages/splm/.

Tobler, W. (1970). A computer movie simulating urban growth in the detroit region. *Economic Geography*, *46*(2), 234–240.

Wink, R. (2014). Regional economic resilience: Policy experiences and issues in Europe. *Raumforschung und Raumordnung*, *72*(2), 83–84.

The manufacturer's authorised representative in the EU is Springer Nature Customer Service Centre GmbH, Europaplatz 3, 69115 Heidelberg, Germany. If you have any concerns regarding our products, please contact ProductSafety@springernature.com

Printed and bound by CPI Group (UK) Ltd, Croydon, CR0 4YY
30/09/2026
02225467-0009